# GLACIAL SYSTEMS AND

# GLACIAL SYSTEMS AND LANDFORMS

## *A Virtual Interactive Experience*

By RYAN C. BELL

ANTHEM PRESS
LONDON · NEW YORK · DELHI

Anthem Press
An imprint of Wimbledon Publishing Company
*www.anthempress.com*

This edition first published in UK and USA 2013
by ANTHEM PRESS
75–76 Blackfriars Road, London SE1 8HA, UK
or PO Box 9779, London SW19 7ZG, UK
and
244 Madison Ave #116, New York, NY 10016, USA

*British Library Cataloguing-in-Publication Data*
A catalogue record for this book is available from the British Library.

*Library of Congress Cataloging-in-Publication Data*
Bell, Ryan C.
  Glacial systems and landforms : a virtual interactive experience /
by Ryan C. Bell.
      pages cm
  Includes bibliographical references and index.
  ISBN 978-0-85728-061-9 (pbk. : alk. paper)
  1. Glacial landforms. 2. Glaciology. I. Title.
  GB581.B445 2013
  551.31–dc23
                                        2013023937

ISBN-13: 978 0 85728 061 9 (Pbk)
ISBN-10: 0 85728 061 9 (Pbk)

Cover photo: The Aletsch glacier © Thinkstock

This title is also available as an eBook.

# CONTENTS

*Acknowledgments*      xi

*List of Figures*      xiii

**How to Use this Book: A Note to the Student and Teacher**      xv

Google Earth Instructions      xvi

    Step 1: Download Google Earth      xvi

    Step 2: Learn to navigate      xvii

    Step 3: Learn to search for locations      xvii

    Step 4: Learn to change the elevation exaggeration      xix

    Step 5: Learn to use the ruler tool      xix

**Chapter 1: Introduction**      1

Chapter 1 Review      6

    Key terms      6

    Concept review and critical thinking      6

    Google Earth analysis      6

**Chapter 2: What is a Glacier?**      9

Definition      9

Mass Balance      9

    Area of accumulation      9

    Why is glacial ice blue?      10

    Area of ablation      11

    Mass balance fluctuations      12

Chapter 2 Review                                          12
  Key terms                                               12
  Concept review and critical thinking                    12
  Google Earth analysis                                   13

## Chapter 3: Types and Locations of Glaciers            15
Topographic Classifications                               15
  Alpine glaciers                                         15
  Continental glaciers                                    17
Temperature Classifications                               19
  Warm glaciers                                           19
  Cold glaciers                                           19
  Polythermal glaciers                                    20
Chapter 3 Review                                          20
  Key terms                                               20
  Concept review and critical thinking                    20
  Google Earth analysis                                   21

## Chapter 4: How Do Glaciers Move?                       23
Mechanisms of Ice Flow                                    23
  Internal deformation                                    23
  Basal sliding                                           24
  Bed deformation                                         24
Rates of Movement                                         24
  Longitudinal movement                                   25
  Transverse movement                                     25
Chapter 4 Review                                          26
  Key terms                                               26
  Concept review and critical thinking                    26
  Google Earth analysis                                   27

## Chapter 5: Ice Structures                              29
Surface Ice Structures                                    29
  Ogives                                                  29
  Crevasses                                               30
Subsurface Ice Structures                                 32
  Layers of accumulation                                  32
  Foliation, folding and faulting                         33

Chapter 5 Review                                    35
    Key terms                                       35
    Concept review and critical thinking            35
    Google Earth analysis                           35

**Chapter 6: Glacial Erosion**                      37
Glacial Debris Entrainment                          37
    Supraglacial sources                            37
    Subglacial sources                              38
Erosional Processes                                 38
    Abrasion                                        39
    Plucking                                        39
    Basal meltwater                                 40
Chapter 6 Review                                    41
    Key terms                                       41
    Concept review and critical thinking            41

**Chapter 7: Landforms of Glacial Erosion**         43
Intermediate-Scale Features of Glacial Erosion      43
    Roches moutonnées                               43
    Whalebacks                                      44
Large-Scale Features of Glacial Erosion             44
    Glacial troughs and fjords                      44
    Hanging valleys                                 46
    Cirques, arêtes and horns                       47
Chapter 7 Review                                    50
    Key terms                                       50
    Concept review and critical thinking            50
    Google Earth analysis                           51

**Chapter 8: Glacial Deposition**                   53
Types of Glacial Deposition                         53
    Direct glacial deposition                       53
    Indirect glacial deposition                     55
Chapter 8 Review                                    56
    Key terms                                       56
    Concept review and critical thinking            56

## Chapter 9: Landforms of Glacial Deposition          57

Landforms Created in an Ice-Marginal Position          57
   Terminal and recessional moraines          57
   Lateral and medial moraines          58
Landforms Created in a Subglacial Position          59
   Drumlins          59
   Eskers          60
Ice Contact Features          60
   Kames          61
   Kettles          61
Chapter 9 Review          61
   Key terms          61
   Concept review and critical thinking          61
   Google Earth analysis          62

## Chapter 10: Ice Ages and Interglacial Periods          63

Natural Causes of Climate Change          63
   Changes in solar output          63
   Changes in the Earth's motions          64
   Plate tectonics          64
   Changes in atmospheric composition          65
How Do Scientists Study Climate Change?          66
   Ice core analysis          66
   Deep-sea cores          67
   Pollen evidence          67
   Atmospheric measurements          67
The Pleistocene Ice Age          68
   Sea-level change          70
   The Great Lakes          70
Chapter 10 Review          73
   Key terms          73
   Concept review and critical thinking          74

## Chapter 11: Periglacial Environments          75

Definition          75
Frost Action          75
   Frost heaving          76
   Thaw weakening          76

Permafrost                                                              77
Periglacial Landforms                                                   79
  Pingos                                                                79
  Thermokarst                                                           80
  Patterned ground                                                      81
Chapter 11 Review                                                       83
  Key terms                                                             83
  Concept review and critical thinking                                  84

**Chapter 12: Glaciers and Global Warming**                             85
Anatomy of the Atmosphere                                               85
The Greenhouse Effect                                                   86
Global Warming                                                          86
  Carbon dioxide                                                        86
  Methane                                                               87
  Nitrous oxide                                                         89
Predicting Climate Change in the Twenty-first Century                   89
  Water-vapor feedback                                                  89
  Cloud-radiation feedback                                              90
  Ocean-circulation feedback                                            90
  Ice-albedo feedback                                                   91
Projected Outcomes of a Warmer World                                    91
  Global warming and alpine glaciers                                    92
  Global warming and continental glaciers                              93
    Greenland                                                           94
    Antarctica                                                          95
  Global warming and rising global sea levels                          96
  Global warming and periglacial environments                          98
Solutions: Applying the Precautionary Principle                        99
Cleanup and Prevention Strategies for Reducing Greenhouse Gas Emissions 100
  Cleanup strategies                                                   100
  Prevention strategies                                                100
International Climate Negotiations                                      101
Individuals Matter                                                      102
Chapter 12 Review                                                      103
  Key terms                                                            103
  Concept review and critical thinking                                 103
  Google Earth analysis                                                104

**Chapter 13: Final Project**                                    105
Assignment                                                      105
Learn to Create a Tour in Google Earth                          106

*Bibliography*                                                  109

*Index*                                                         111

# ACKNOWLEDGMENTS

Firstly I wish to thank Tej Sood, Rob Reddick and their team at Anthem Press for their advice and guidance during the preparation of this book. I am equally indebted to a diverse group of readers who brought an original perspective and unflinching insights; this includes John Banker, Jeanne Kaidy, David King, Shawn McNamara and Sean Metz. I would also like to acknowledge the contributors to the content and images of this book, many of whom I do not know personally, but whose works I have read and studied. Many of these works are cited throughout this book.

I am also grateful to my colleagues, past and present, at Pittsford Sutherland High School for their inspiration and companionship. In particular I am deeply indebted to Georgiann Francione for introducing me to the wonders of Earth science and perhaps more importantly, for showing me how to teach it. Also, thanks to Sandy Stein for showing me that there is more to teaching than just content. Most importantly, thank you to my students whose enthusiasm constantly allowed me to look at glaciers with a fresh perspective. Unfortunately, there is not enough space to acknowledge them all, but without the whole lot of you – past, present and future – this book would not have been possible.

Heartfelt thanks to my friends and family for their support and encouragement; I am especially grateful to my brothers Brandon and Shannon Bell for their interest and words of encouragement during the writing process. I am most obviously indebted to my parents. My gratitude to them is boundless, as they have supported me in more ways than I am able to count. Finally, thank you to Jacqueline Bell, my wife, without whom my world would not be possible.

# LIST OF FIGURES

| | | |
|---|---|---|
| Figure 1 | Larsen-B ice shelf breakup | xvi |
| Figure 2 | Google Earth's "Fly To" box | xviii |
| Figure 3 | Google Earth "Options" | xix |
| Figure 4 | Adjusting "Elevation Exaggeration" in Google Earth | xx |
| Figure 5 | Google Earth's ruler tool | xx |
| Figure 6 | Glacier ice formation | 10 |
| Figure 7 | Diagram of a glacier showing components of mass balance | 11 |
| Figure 8 | Alpine glacier – Mount Everest | 16 |
| Figure 9 | A tidewater glacier along the South Georgia Island's southeast coast | 17 |
| Figure 10 | Continental glacier – Antarctica | 18 |
| Figure 11 | Measuring snowpack in a crevasse on the Easton Glacier, Mount Baker | 30 |
| Figure 12 | The Khumbu Icefall, Mount Everest | 31 |
| Figure 13 | Crevasses and icefalls as seen from space on an unnamed glacier in the Himalayan mountain range | 32 |
| Figure 14 | Diagram of glacial plucking and abrasion | 40 |
| Figure 15 | A roche moutonnée in the Cadair Idris Valley, Snowdonia, Wales | 44 |
| Figure 16 | Glacial trough or U-shaped valley in Mount Baker Snoqualmie National Forest | 45 |
| Figure 17 | Glacial trough or U-shaped valley in the Swiss Alps | 46 |
| Figure 18 | Geiranger Fjord, Norway | 47 |
| Figure 19 | Fjords found along the Southern Patagonian Ice Field | 48 |
| Figure 20 | The Garden Wall, an arête in Glacier National Park | 48 |
| Figure 21 | Glacial tarn formation | 49 |

Figure 22   The Matterhorn                                                    49
Figure 23   Angular glacial erratic on Lembert Dome in Yosemite
            National Park                                                     54
Figure 24   Glacial moraines in Alberta, Canada                              58
Figure 25   A medial moriaine in the Upsala Glacier found in
            the Southern Patagonian Ice Field                                59
Figure 26   Drumlin in Clew Bay, Ireland                                     60
Figure 27   Changes in Earth's axial tilt                                    65
Figure 28   An artist's rendition of the Earth during an ice age             69
Figure 29   The Great Lakes                                                  71
Figure 30   Pingos near Tuktoyaktuk, Northwest Territories, Canada           80
Figure 31   Thermokarst lakes on Alaska's North Slope region                 81
Figure 32   Patterned ground structures on the Svalbard Archipelago          82
Figure 33   Deforestation in the Amazon interior from 1992 to 2006           88
Figure 34   Alpine glacial recession: The retreat of the Gangotri
            Glacier in India                                                 92
Figure 35   An enormous piece of ice, roughly 250km² in size,
            breaking off from the Petermann Glacier along the
            northwestern coast of Greenland in 2010                          93
Figure 36   Mass balance atmospheric circulation                             95
Figure 37   Creating a new folder in Google Earth                           106
Figure 38   Adding a placemark in Google Earth                              107
Figure 39   Adding a title to a new placemark in Google Earth               107
Figure 40   Arranging a tour folder in Google Earth                         108
Figure 41   Playing a tour in Google Earth                                  108

# HOW TO USE THIS BOOK: A NOTE TO THE STUDENT AND TEACHER

Glaciers today are more accessible than ever. The proliferation of high-resolution satellite imagery in recent years has allowed scientists to investigate even the most remote and inhospitable glacial environments from the comfort of their own living rooms. In fact, anyone with an Internet connection and an interest can freely use Google Earth to monitor and study almost any glacier in the world.

Granted field research is, and probably always will be, an invaluable component of glaciology. However, one no longer needs to be an experienced mountaineer or hardened polar adventurer to learn about glacial systems and landforms. Data from satellite imagery has refined our understanding of not only the extent of glacial ice, but also the processes that govern glacial landscapes. Thus, many scientists are trading in their skills with an ice axe for a new skill-set equipped to analyze and interpret satellite images. These skills are becoming increasingly important as human activities continue to transform the face of the Earth. Scientists and environmental groups, for example, rely on sophisticated satellite imagery to monitor and study the effects of human-induced climate change. In fact, many scientists reported that NASA's satellite imagery of the 2002 collapse and subsequent disintegration of the Antarctic Peninsula's Larsen-B ice shelf captured global warming in motion (Figure 1).

Included in this book is a series of high-resolution satellite images from NASA's Earth Observatory; these spectacular images highlight specific landforms indicative of glacial activity. By studying these images, the reader will not only start to recognize the patterns and processes commonly found within glacial landscapes, but will also develop skills in map analysis and interpretation. To further build these skills, several chapters end with a set of latitude and longitude coordinates. Using Google Earth, the reader can "fly to" these locations and investigate a variety of glacial landforms across the planet. In this way, the book

**Figure 1.** Larsen–B ice shelf breakup

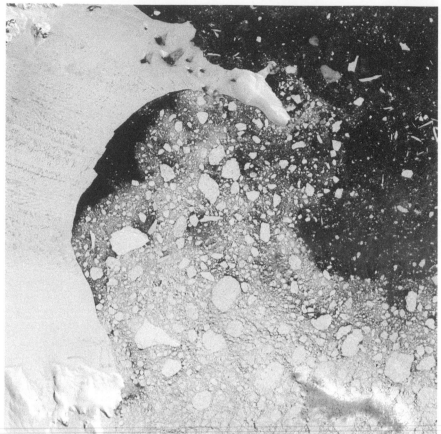

Source: NASA.

offers a virtual interactive experience in which the reader can visit and explore glacier environments in 3D.

# Google Earth Instructions

### Step 1: Download Google Earth

In order to use this program, you must have high-speed Internet and the Google Earth software installed on your computer. This program is free to download and can be found at http://earth.google.com. Helpful video tutorials can be found at http://www.google.com/earth/learn/. Additional help and technical support may be found at http://support.google.com/earth/?hl=en.[1]

---

1  These, and all other links mentioned in this book, were active at the time of publication.

## Step 2: Learn to navigate

Once downloaded, play around with it by flying to areas of interest. It is important that you familiarize yourself with the navigation tools found in the top-right corner of your 3D viewer. If for some reason your navigation tools are not appearing on your screen, you can set them to appear automatically by clicking "View" found along the top main-menu bar and selecting "Show Navigation."

The navigation controls are divided into three components. The bottom control, known as the "Zoom Slider" allows you to zoom in and zoom out. By dragging this slider or clicking on the icons at either end of the slider, you will be brought closer or farther away from the center of your view. The same can be done by double clicking on any location in the 3D-view screen as well. To experience a different angular perspective, try zooming all the way into a location. Once you are completely zoomed in, continue to click the "zoom in" icon and Google Earth will tilt your viewing angle so that you can view the horizon from the ground. The second navigation control located directly above the "Zoom Slider" is the "Move Joystick"; by dragging the hand in the center toward a desired direction, the "Move Joystick" allows you to fly around. Finally, the third component located at the top of the navigation controls is the "Look Joystick," which adjusts your perspective as though you were turning your head while standing in the same spot.

If you do not have a mouse or would rather use your keyboard, all of the navigation controls discussed above can be accessed with the following key strokes.

| Navigation | Keyboard commands |
| --- | --- |
| Right, Left, Up and Down | Arrow Keys |
| Zoom In/Zoom Out | Plus Key/Minus Key |
| Change Angular Perspective | Shift and Up Arrow/Shift and Down Arrow |
| Rotate Clockwise/Counter Clockwise | Shift and Left Arrow/Shift and Right Arrow |

These navigational tools will help you view important landforms from a variety of perspectives.

## Step 3: Learn to search for locations

At the end of several chapters in this book you will find a series of latitude and longitude coordinates; these will direct you to a variety of glacial landforms.

Some exercises will identify the locations for you, while others will ask you to interpret and identify what you observe. In order to locate these positions, type the latitude and longitude coordinates into the "Fly To" box under the search panel and click on the magnifying glass.

**Figure 2.** Google Earth's "Fly To" box

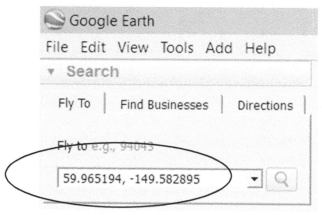

Source: Google Earth is a trademark of Google, Inc.

You will notice that the format of the latitude and longitude coordinates does not include the cardinal directions north, south, east or west. Rather, the coordinate system used in this book is presented in decimal degrees. Instead of directions indicating which hemisphere a location is in, the decimal-degree format uses positive and negative values. Positive latitude values indicate locations in the Northern Hemisphere, and negative latitude values indicate positions in the Southern Hemisphere. Similarly, positive longitude values represent the Eastern Hemisphere, while negative values represent the Western Hemisphere. For example, a location in Alaska has a latitude and longitude decimal-degree coordinate of 59.965194, −149.582895. The positive latitude value and negative longitude value tells us that the location is in the northwestern hemisphere.

Many activities in this book will give you specific latitude/longitude coordinates to fly to, however, occasionally you might be asked to identify the coordinates of a location; to do this, simply move your cursor over the location desired. At the bottom of your 3D-viewer screen you will see a slightly shaded banner, which indicates not only the latitude/longitude coordinates of the location but also its elevation. At the start of your activity, make sure these values are presented in the decimal-degree format and the elevation is reported in meters. To change these settings, click on the "Tools" icon found along the top of the screen in the main menu. Select "Options" and make the appropriate changes as indicated by the screen capture in Figure 3.

**Figure 3.** Google Earth "Options"

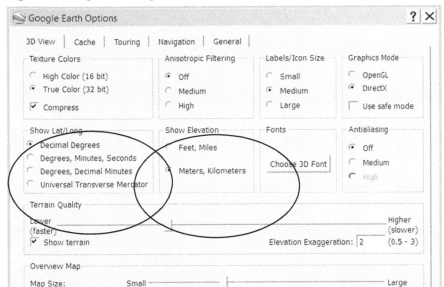

Source: Google Earth is a trademark of Google, Inc.

## Step 4: Learn to change the elevation exaggeration

While at your location you may want to make some adjustments for better viewing. The elevation exaggeration, for example, can be changed by selecting "Tools" from the main menu found along the top of the screen. Next, open the options dialogue box by selecting "Options." Toward the bottom of the options dialogue box you will see a box titled "Elevation Exaggeration" (see Figure 4 on the next page). You may type in a range of values from 0 to 3. Higher values will increase the relief of the landscape, while lower values will decrease the relief. Increasing the relief to a value of 2 or 3 might be desirable in a relatively flat landscape; however, this exaggeration is often a bit overwhelming and unrealistic in highland environments.

## Step 5: Learn to use the ruler tool

Another useful feature within Google Earth is the ruler tool. Located in the toolbar at the top of the Google Earth window, the ruler tool allows you to measure the length of a straight line or path. To find the distance of a straight

**Figure 4.** Adjusting "Elevation Exaggeration" in Google Earth

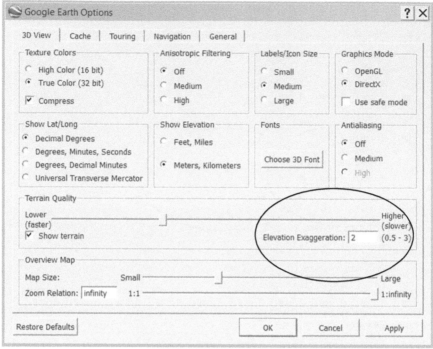

Source: Google Earth is a trademark of Google, Inc.

line, simply designate the two end points by left clicking two points with your mouse. Some activities in this book may ask you to measure the distance, for example, between a glacier's head or starting point and the snout of a glacier or terminus.

**Figure 5.** Google Earth's ruler tool

Source: Google Earth is a trademark of Google, Inc.[2]

---

2 These and all other screenshots from Google Earth were accurate at the time of publication.

# 1

# INTRODUCTION

In 1965, scientist James Lovelock formulated a hypothesis that revolutionized the way we think about the Earth. Lovelock postulated that the Earth is a living, self-regulating organism capable of monitoring and adjusting its climate and chemical composition (1979). While thought provoking, the idea was initially ignored because without sufficient scientific evidence, it lacked teeth. Years later, following exhaustive research and extensive mathematical modeling this hypothesis morphed into what Lovelock called **Gaia theory**. Named after the Greek goddess of Earth, essentially this theory posits that Gaia is the sum of all Earth's parts working together to achieve relatively constant conditions through self-regulation. In other words, the oceans, life, air and surface rocks are organs or parts of an entity that operate collectively in an attempt to maintain a sort of planetary homeostasis and provide optimal physical and chemical conditions for life on Earth (Lovelock 1979, 10).

Although Gaia theory has provided a new paradigm and framework in which to study the Earth, it has not been fully embraced by the scientific community. In fact, many scientists dismiss this theory entirely due to its religious and spiritual overtones. Despite these unfounded criticisms, most scientists accept the notion that there are strong interrelationships between the biotic (living) and abiotic (nonliving) environment. Consequently, Earth system science has replaced the name Gaia in an effort to make it easier for the scientific community to digest. The principals, however, remain the same.

Under this new lens, a **system** can be defined as a set of components that function and interact in some regular way within a defined boundary. Earth's systems can be divided into four major interdependent parts known as "spheres." These spheres include the **lithosphere** (rock, soil and sediment), **hydrosphere** (water), **atmosphere** (air) and **biosphere** (living things). One additional sphere that cannot be ignored when describing the coordinated efforts to achieve planetary homeostasis is the **cryosphere**. The cryosphere represents the ice-covered regions of the planet. This may include snow, lake ice, sea ice, ice caps, ice sheets, permafrost and glaciers. For a variety of reasons, glaciers are a critical component of this vital Earth system.

Glaciers exist in a balance. Like most systems, glaciers have inputs from the environment, flows of matter and energy within the system and outputs to the environment. Alpine glaciers, for example, exist between a balance of snowfall at high elevation and snowmelt at low elevation. Many glaciologists however, argue that this balance has recently been knocked off kilter. A growing body of evidence suggests that the cause is human-induced climate change. Projections are that global surface temperatures could rise 1.5–6°C over the next century, with significant regional variation. Systems typically respond to such changes through a series of **feedback loops**, a response that increases or decreases the change to the system. Corrective measures known as **negative feedback loops**, for example, reverse the direction of change in a system. Negative feedbacks suppress change and tend to maintain stability in a system. Many scientists explain negative feedback loops by comparing them to a programmable thermostat in a home heated by a furnace. When the temperature of a home rises to a certain temperature, the thermostat is set to turn off the furnace and the house begins to cool. In effect, the cryosphere is the world's thermostat. Despite fluctuations in solar output over the past several thousand years, the Earth's average temperature has remained relatively constant and favorable for life, in part because glacial ice helps to cool the Earth by reflecting incoming sunlight.

However, not all changes in systems, such as those that affect glaciers are corrective. Systems also respond to change with positive feedback. A **positive feedback loop** causes a system to change further in the same direction. These changes often destabilize a system. As global temperatures rise, for example, glacial and polar ice melts at accelerating rates. Positive feedbacks emerge as the melting ice and snow exposes much darker land and sea areas, which reduces Earth's reflectivity. Thus, more solar energy is absorbed into the Earth's surface, which in turn warms the atmosphere. In a vicious cycle, more ice melts and temperatures continue to rise. In this way, positive feedback loops are often self-perpetuating and destructive.

Unfortunately, these positive feedback loops have escalated in recent years. The industrial revolutions of the eighteenth and nineteenth centuries instigated dramatic transformations within Earth's atmosphere. As populations continue to industrialize during the twenty-first century, humans are increasing atmospheric concentrations of greenhouse gases such as carbon dioxide through the combustion of fossil fuels and the clearing of forests. These gases trap outgoing heat energy, which in turn raises global temperatures. In some parts of the world, warmer temperatures are melting glaciers at alarming rates. Recent surveys indicate that 90 percent of the ice shelves on the Antarctic Peninsula are retreating 50m per year. Likewise, the ice caps located in Greenland are disappearing twice as fast as they were a few years ago. Moreover, alpine glaciers are vanishing right before our eyes. For example, it is expected that the permanent ice found on top of Mt Kilimanjaro, will be gone by the year 2015. Similarly, researchers in the US government have calculated that there will be no more glaciers in Montana's Glacier National Park by the year 2030. While the realities of global warming are currently being made evident by the cryosphere, the future implications will be far more reaching.

The paragraph above reads a bit like a Doomsday report. However, it was not written in an attempt to fix you in a state of fear, but rather to pose some important questions. For instance, why should we care if there is less ice in Glacier National Park? To this end, why should we care if all of the world's glaciers waste away? At first glance, it appears that these often-remote inhospitable landscapes are disconnected from our lives. After all, most of us will never have the opportunity to visit the ice sheets of Antarctica or the highland ice fields of Patagonia. These formidable landscapes are harsh and can be unforgiving to the careless hiker. For these reasons glaciers are commonly out of sight, out of mind.

Despite the physical distances, mental disconnections and the hazards of glacial travel, many scientists have braved glacial terrain and journeyed through significant portions of our icy planet. These explorations have contributed to a large and varied body of knowledge that has reshaped our understanding of Earth's systems. While this field research is essential, one does not need to be a mountaineer or a hardened polar adventurer to understand what makes a glacier tick. The explosion of satellite imagery and remote-sensing technologies in recent years has further helped reveal the hidden truths of glacial environments by providing new methods of mapping and measuring glacial ice.

It turns out, for example, that the cryosphere, or portion of the Earth that is dominated by ice, is extensive. Glacier ice covers 10 percent of the Earth's land surface, and this ice can be found on all seven continents spanning an area just less than $16,000,000km^2$. Until recently, geologically speaking, ice covered up to 30 percent of the land during the last ice age. This deep freeze came to a close 10,000 years ago and has placed the Earth into what scientists call an interglacial period ever since. These glacial and interglacial transitions are not new or uncommon; large sheets of ice have advanced and retreated across

North America and Europe several times over the past 2 million years. The truth is we are living in the midst of a glacial era.

Maybe the most important truth, however, is the fact that past and present glaciations profoundly influence almost every aspect of life on Earth. Globally, glaciers affect sea level, control the water cycle, drive ocean currents and enhance atmospheric circulation. These global climatic factors help determine the geographic distribution of soils, plants and animals. Earth's biodiversity is in large part a product of many glacial periods. This may seem a bit counterintuitive considering that some of the largest mass extinctions in Earth's history accompany periods of glaciation. However, evolutionary biology teaches us that the proliferation of life often follows the loss of it. In a process known as adaptive radiation, the vacant ecological niches created from extinction are quickly (geologically speaking) filled through a process known as speciation.

Humans tend to benefit from glaciers as well. Many human civilizations, for example, depend upon the water resources from mountain glaciers. Glacial meltwater in the high Andes provides drinking water and irrigation to millions of people in tropical countries such as Bolivia and Peru. Likewise, a complex irrigation system supplies agricultural communities living in the rain shadow of the Swiss Alps with an abundance of water. The mineral-laden meltwater also provides critical nutrients for what would normally be an unproductive landscape.

In addition to the present-day glaciers of the world, ancient glaciers that have long since melted are currently providing vital water resources to people across the planet. Glacially carved lakes such as the Finger Lakes and the Great Lakes of North America sustain large populations with their vast reservoirs of freshwater. While these surface-water deposits of ancient ice sheets are well known, perhaps less obvious are the massive subsurface repositories of freshwater created by glacial action and meltwater. Such groundwater supplies are stored in porous, water-saturated layers of sand, gravel or bedrock known as aquifers. One of the largest known aquifers in the world is North America's Ogallala Aquifer. Deposited during the last ice age over 15,000 years ago, the water withdrawals from the Ogallala Aquifer have transformed the arid high plains of the Midwest into one of the most productive agricultural regions in the United States.

Hydroelectric power generation is another added benefit of glacial meltwater. Alpine countries such as Norway, Austria and Switzerland utilize dams to retain and store runoff from ice melt when electrical demands are low. When the demand increases, the water is released to spin massive turbines that generate electricity. This renewable source of energy is not without its environmental impacts, but it does offer an alternative to the often-high environmental consequences of burning fossil fuels.

Aside from their instrumental values, glaciers are aesthetically magnificent. If the enormity and striking bluish color of glacial ice does not command one's attention, the sculpted landforms produced from the powerful forces of glacial erosion will leave a visitor breathless. The hanging valleys plunging with spectacular waterfalls along the fjords of Alaska's southeast coast are just one example of countless transcendent landscapes left behind in the wake of glaciers in retreat. Astounded by the beauty of Alaska's Glacier Bay, naturalist, John Muir, wrote:

> The sky to-day is mostly clear, with just clouds enough hovering about the mountains to show them to best advantage as they stretch onward in sustained grandeur like two separate and distinct ranges, each mountain with its glaciers and clouds and fine sculpture glowing bright in smooth, graded light. Only a few of them exceed five thousand feet in height; but as one naturally associates great height with ice-and-snow-laden mountains and with glacial sculpture so pronounced, they seem much higher.
>
> *Travels in Alaska* (1989 [1915], 209)

This scenic value is big business for many communities living in the midst of past and present glacial environments; these landscapes are major tourist attractions in many parts of the world. From adventurous alpine skiing in the Swiss Alps to leisurely wine tasting along the Finger Lakes of central New York – glacial landscapes generate billions of dollars annually.

Clearly, glaciers today, and in the past, have had a profound influence on every facet of the planet. The intimate relationship between ice and life in particular is significant, yet it is poised on a knife's edge. Ice is an extremely fragile substance. The slightest change in environmental conditions such as temperature, pressure and salinity can move ice from a solid to a liquid. Unfortunately, as noted earlier, human activities are triggering these transformations at alarming rates and many of the world's glaciers are melting.

As the world population grapples with global climate change, the consequences, answers and solutions will be found in glacial ice. "Modern glacial environments," according to scientist John Menzies, "hold an essential key to our knowledge of the present, as well as past and future global environmental conditions" (1995, 1). It is no longer enough to restrict the wonders of these majestic environments to a select group of professional scientists. Broad public support is required to prevent the breakup and disintegration of Earth's cryosphere. The purpose of understanding glacial systems and landforms has therefore transformed itself from an area of scientific interest to one of necessity and one fundamental to present and future generations.

# Chapter 1 Review

### Key terms

Gaia theory

System

Feedback loop

Positive feedback loop

Negative feedback loop

Lithosphere

Hydrosphere

Atmosphere

Cryosphere

### Concept review and critical thinking

1. What is a system? Explain why a glacier is considered a system.

2. What is a feedback loop? Distinguish between a positive feedback loop and a negative feedback loop in a system and provide an example of each as they pertain to glaciers.

3. Why are glaciers important? Describe two economic and two ecological benefits of glaciers.

4. Evaluate Gaia theory. Is it science or is it pseudoscience? Explain.

### Google Earth analysis

John Muir, one of America's most prolific ecological thinkers, first visited Alaska's Glacier Bay in 1879. Fascinated by the blossoming science of glaciology, Muir wanted to witness glaciers in action and substantiate his theories that the hanging valleys and U-shaped valleys found in his beloved Yosemite Valley, California were in fact residual products of glacial erosion. While exploring the glacially carved, long, narrow inlets of Glacier Bay, he confirmed his suspicions and was immediately struck by the beauty of the dramatically sculpted mountainous terrain. The active glaciers found within Glacier Bay helped Muir gain insight into the ancient glaciers that once occupied Yosemite Valley.

## Instructions

Take a trip to Glacier Bay, Alaska using Google Earth. Type in the latitude/longitude coordinates found below into the "Fly To" box. Practice navigating the area and changing angular perspectives. Locate the terminus (where the glacier meets the water) of two glaciers and follow the glaciers from where they end to where they originate in the mountains. Make five general observations of the landscape.

*Glacier Bay, Alaska: 58.695460, −136.184992*

Next, take a trip to Yosemite Valley, California using Google Earth. Type in the latitude/longitude coordinates found below into the "Fly To" box. Fly through the valleys and make five general observations of the landscape. Compare your observations of Yosemite to your notes of Glacier Bay. Do you see any similarities? Do you think Muir's reasoning was correct in saying that like Glacier Bay, Yosemite Valley was formed in part by glaciers? Justify your answer.

*Yosemite Valley, California: 37.743304, −119.575871*

**2**

# WHAT IS A GLACIER?

## Definition

Many scientists define a glacier as a perennial mass of land-based ice that forms from the recrystallization of snow. This simplistic definition, however, does not tell the whole story. A glacier is more than just an ever-present chunk of ice. Glaciers transport and deposit ice and sediments, and in doing so, they scour the landscape and shape the Earth. Lifelike, glaciers expand and contract; they advance and retreat. Simply put, glaciers are dynamic.

## Mass Balance

These dynamics are essential to our understanding of glacial formation. As pointed out in the introduction, glaciers are systems with both inputs and outputs. Scientists refer to the gain and loss of ice in a glacier system as the **mass balance**; this balance manifests itself in a glacier's **equilibrium line**. It is helpful to think of the equilibrium line as a budget. Just as a monetary budget would take into account earning and spending, a glacial budget aims to quantify snowfall and snowmelt. The equilibrium line therefore represents the interface between a glacier's two principal zones. Glaciologists refer to these two zones as the **area of accumulation** (snowfall area) and the **area of ablation** (melting area).

### Area of accumulation

The area of accumulation typically occurs at high elevations above the equilibrium line where snowfall is great enough to last throughout the following summer. Snowfall is the primary source of accumulation and therefore a major contributor to the formation of glacial ice. Each year, winter snow is deposited on the previous year's snowfall; with each subsequent deposition, more weight compresses

**Figure 6.** Glacier ice formation

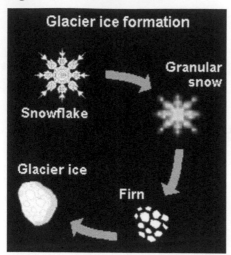

Source: Luis Maria Benitez.

the underlying snow. This process initiates a sequence of events that lead to the formation of **glacial firn** and then finally glacial ice. Glacial firn is the transitional product of glacial-ice formation. Light fluffy snow falls onto the zone of accumulation and the original intergrown crystalline structure of the snowflake quickly condenses under the weight of overlying snow. After a year, compaction coupled with localized melting and refreezing, transforms the once intricate hexagonal structure of a snowflake into homogenous granular ice known as firn (Sugden and John 1976). The rounded grains of the year-old snow or firn continue to increase in hardness and density as snowfall thickens the overburden and adds to the overlying weight. This pressure is further intensified by the tug of gravity when the glacier flows as a collective unit downhill. The transformation from snow to glacier ice continues. The firn recrystallizes and grows larger, reducing the volume of air-filled pores. The state of glacier ice is reached when the pore spaces between firn have been sealed, which ultimately reduces the permeability of the ice (Sugden and John 1976).

## Why is glacial ice blue?

The sequence of the aforementioned events also yields one of the most notable characteristics of glaciers – their bluish color. Many have observed that snowflakes appear white. In fact most ice on planet Earth is colorless or white. So why does glacial ice often appear blue? The answer resides in the reflectivity and absorption of different wavelengths of light. Visible light consists of a continuum of different colors broken down by wavelength; the color red is found at the long end of the light spectrum while blue occurs at the short end of the spectrum. Objects that appear white, such as ice and snow under normal conditions, reflect all wavelengths of light with equal intensity. Glacial ice, however, is under enormous amounts of pressure. Therefore, the density of glacial ice is much greater. As ice density increases, the amount of shorter wavelength light reflected increases. This means that glacial ice absorbs more light at the red end of the spectrum and the human eye detects the light reflected at the blue end of the spectrum. Thus, glaciers appear blue.

**Figure 7.** Diagram of a glacier showing components of mass balance

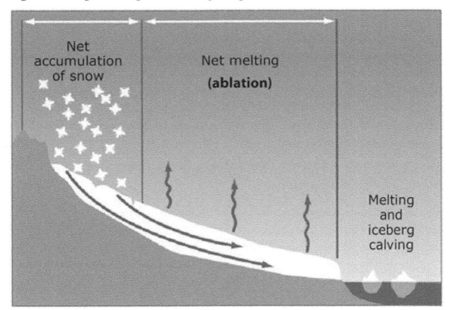

Source: USGS.

## Area of ablation

The ablation area refers to the glacial zone located below the equilibrium line where there is a net loss of ice. Ablation, also known as **wastage**, refers to the processes by which snow and ice are lost from a glacier. Ablation occurs from two primary processes – melting and **calving**. Melting is the principal process controlling wastage (Sharp 1988). While melting is directly related to solar radiation, the extent of melting is controlled by a variety of interconnected factors. One such factor controlling the rate of melting is a glacier's **albedo**, or measurement of surface reflectivity. Generally, light-colored surfaces reflect more sunlight than dark-colored surfaces. Fresh snowfall, for example, reflects 80–95 percent of the incoming solar radiation thereby reducing ablation. Melting ice and snow found in the area of ablation, however, is often dirty with dark-colored sediments and therefore absorbs more solar radiation. Where albedo is reduced, melting often accelerates.

Other factors controlling the rate of melting include water vapor and wind. High levels of moisture in the atmosphere are conducive to wastage. As water vapor condenses into liquid water it releases a significant amount of energy. Moreover, low-lying clouds and fog insulate the air over a glacier by reflecting and reradiating Earth's outgoing infrared energy. "The best of all worlds," according to the geologist Robert Sharp, "in terms of wastage, is to have warm air heavily laden with water vapor blowing continually across a glacier" (1988, 12).

Calving is another mechanism of wastage. For some of the largest glaciers in the world, calving is their primary form of ablation. Calving occurs when glaciers terminate at the sea or share a boundary with a body of water. At the glacial margins, large blocks of ice detach and fall into the water. The Antarctic ice sheet, for example, loses much of its mass through calving. The massive tabular icebergs produced from such wastage can be up to five hundred meters thick and hundreds of kilometers in length.

## Mass balance fluctuations

The mass balance of any particular glacier is subject to change. Just as a monetary budget prospers from a surplus or suffers during deficit, glaciers thrive and flounder. Consequently, the equilibrium line fluctuates in elevation accordingly. Equilibrium lines, for example, move down the ice when accumulation exceeds ablation. During such periods of higher snowfall and reduced wastage, glaciers are said to advance. Conversely, equilibrium lines move to higher elevations when ablation exceeds accumulation. During such periods of rising surface temperatures and reductions in snowfall, glaciers are said to retreat. These changes, although not instantaneous, are responses to changing environmental conditions over time. We will explore these changes later in the text.

# Chapter 2 Review

### Key terms

| | |
|---|---|
| Mass balance | Glacial firn |
| Equilibrium line | Wastage |
| Area of accumulation | Calving |
| Area of ablation | Albedo |

## Concept review and critical thinking

1. Distinguish between the area of accumulation and the area of ablation.

2. Explain how the mass balance of a glacier relates to a monetary budget.

3. Describe how glacial ice forms.

4. Why does glacial ice appear blue?

5. Identify two mechanisms in which wastage occurs.

## Google Earth analysis

Alaska's Juneau Icefield Research Program has monitored the mass balance of the Mendenhall Glacier since 1946. From 1946 to 2005 the terminus of the Mendenhall Glacier has retreated 580m. This means that ablation or snowmelt has exceeded snow accumulation. You can often see the equilibrium line on alpine glaciers because it separates the new, fresh snow found in the accumulation area from the old, dirty snow found in the area of ablation.

## Instructions

Take a trip to Alaska's Mendenhall Glacier using Google Earth. Type in the latitude/longitude coordinates found below into the "Fly To" box.

*Mendenhall Glacier: 58.457201, −134.536041*

1. Examine the terminus of the Mendenhall Glacier. What method of wastage dominates?

2. Locate the equilibrium line separating the accumulation area from the ablation area. Measure the length of the ablation area using the "Ruler Tool." What is the distance from the terminus to the equilibrium line?

3. Calculate the rate of melting for the years 1946–2005.

$$\text{Melting rate} = \frac{\text{Distance of retreat}}{\text{Years}}$$

4. If the rate of melting remains constant, how many years will it take for the glacier to recede to the equilibrium line?

# 3

# TYPES AND LOCATIONS OF GLACIERS

Classifying glaciers is complicated. "The form glaciers take," according to geologists Douglas Benn and David Evans, "is a function of climate and topography, and the morphology of any one glacier is unique to its location on the Earth's surface" (2010, 5). Consequently, no universal classification system exists. Nevertheless, to make description and study easier, scientists employ a variety of strategies, organizing glaciers by size, shape, internal temperature and underlying topography. Using the lens of topography, glaciers can be separated into two major groups – **alpine glaciers** and **continental glaciers**.

## Topographic Classifications

### Alpine glaciers

Alpine glaciers, also known as mountain glaciers or valley glaciers, form at higher altitudes on the slopes of mountains. Confined by the underlying topography, the ice stays within valley walls, forming a large river of ice and snow that moves slowly downhill under the influence of gravity. These mountain glaciers exist at all latitudes and are found above the **snowline**, which is the lowest elevation where temperatures remain cold enough all year to promote snowfall accumulation and glacial ice formation. As one travels from the equator to the poles, he or she might notice a dramatic difference in the snowline's elevation. Such latitudinal variations in snowline elevations are a function of differences in solar radiation. Snowlines in equatorial regions, for instance, are higher in elevation because these latitudes receive a higher angle and intensity of solar radiation. Conversely, the snowlines found in polar regions exist on the lower slopes of mountains in response to lower solar intensities. In fact, many high-latitude valley glaciers terminate at sea level. These **tidewater glaciers** found in areas like the coasts and fjords of Alaska and Greenland produce spectacular sights as their decay calves into the sea (see Figure 9).

**Figure 8.** Alpine glacier – Mount Everest

Source: NASA.

Mountain glaciers vary in size. The largest alpine glaciers are **highland ice fields**; these vast sheets of ice encapsulate many square kilometers in polar and subpolar regions. Unlike the ice sheets that blanket Antarctica and Greenland, highland ice fields lack a domed-like surface and are strongly influenced by the underlying topography (Sugden and John 1976). Mirroring the valleys and ridges below the glacier, the ice possesses dramatic undulations in its relief. Occasionally, mountain peaks surpass the glacier's thickness and breach the ice surface; these isolated peaks and ridges are known as **nunataks**. Renowned for their highland ice fields, they include the remote locations of the Canadian Arctic, southeast Alaska, Patagonia and the Antarctic Peninsula.

Highland ice fields often serve as source zones for valley glaciers. Outlet glaciers discharge vast quantities of ice from these accumulation areas into deep bedrock valleys below. The tongues of valley glaciers can measure tens of kilometers in length and over a thousand meters deep. Despite their painfully slow pace (perhaps a fraction of a meter per day), their momentum can plunge a glacier's leading edge well below the snowline. Other source zones for these outlet glaciers include ice caps, ice sheets and theater-like basins known as cirques (Hambrey and Alean 2004).

Where alpine glaciers escape the narrow confines of their valley walls and flow out onto a level or nearly level plain, they form lobed-shape glaciers commonly referred to as **piedmont glaciers**.

**Figure 9.** A tidewater glacier along the South Georgia Island's southeast coast

Source: NASA.

The smallest alpine glaciers include **ice aprons, ice fringes** and **glacierets**. These accumulations of thin snow and ice are distinguished from one another by their topographic setting. Ice aprons can be found clinging to the sides of mountains, ice fringes occupy small depressions along coasts and glacierets, formed from the accumulation of drifting snow or avalanches are commonly found on terrain with less dramatic relief. These relatively small glaciers can be distinguished from large patches of snow because of the significant movement that these ice masses undergo.

## Continental glaciers

Continental ice sheets are the largest glaciers on Earth. Unconstrained by the subglacial topography, ice sheets cover vast areas of land to considerable depth. Continental glaciers originate from a central region of accumulation and radiate outward. It is helpful to think of the formation of a continental glacier as the creation of a pancake. The batter is poured onto the cooking sheet at a central point and as the batter piles up, the pancake begins to spread outward. In the same way, ice domes build up symmetrically over a central land area. The majority of the ice is found close to the center where ice thickness commonly exceeds 3,000m. The convex dome of these ice sheets gently slopes from the

**Figure 10.** Continental glacier – Antarctica

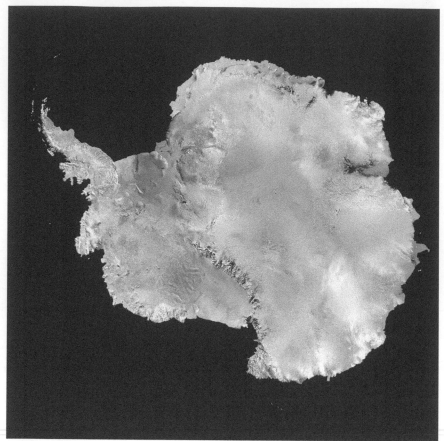

Source: NASA.

interior and steepens along the edges where wastage occurs. Consequently, the interior regions of ice sheets move relatively slow in comparison to the faster-moving ice zones found along the ice sheets margins. Ice streams and outlet glaciers discharge most of the ice toward the continental glacier's periphery. Many coastal outlet glaciers flow offshore where the accumulation of ice and snow produces extensive **ice shelves**. These gently sloped floating glacier tongues range in thickness from 200m to over two kilometers and frequently calve into the sea, yielding massive icebergs that have been measured over a hundred kilometers across (Hambrey and Alean 2004, 14).

There are only two ice sheets in the world. One covers most of Greenland near the Arctic Circle and the other one covers the Antarctic continent at the

South Pole. Together, these two ice masses cover an area of 14.1 million km$^2$, and contain a volume of ice that if melted would raise global sea levels by 70m. To put it mildly, these ice sheets are massive.

To be considered a continental ice sheet, a glacier needs to exceed an area of 50,000km$^2$. Continental glaciers below this minimum size are called ice caps. Notwithstanding size, ice caps are similar to ice sheets in every respect. Geographically, ice caps are found in the polar and subpolar regions of Iceland, Baffin Island, Ellesmere Island and the High Arctic archipelago of Svalbard.

# Temperature Classifications

Another way to classify a glacier is by taking its temperature. Glaciologists determine the internal temperature of a glacier by lowering probes deep into boreholes at the surface. These measurements have revealed that there are both warm and cold glaciers.

### Warm glaciers

A **warm glacier** seems paradoxical. Relatively speaking, all glaciers are cold, however, the internal temperatures within and amongst glaciers can vary significantly. Warm glaciers or temperate glaciers are those that exist at the pressure-melting temperature of the ice throughout. Since water freezes at lower temperatures as pressure increases, a warm glacier's temperature decreases with depth. Hovering at the melting point also means that liquid water can exist from the surface to its base. Basal melting is especially common within warm glaciers as rock obstacles may exert additional pressure and friction, which may cause ice to melt. Consequently, perennial meltwater is a good indicator of a temperate glacier (Hambrey and Alean 2004). Warm glaciers are commonly found in mountain regions outside of the Arctic and Antarctic.

### Cold glaciers

Conversely, **cold glaciers** have temperatures below the pressure-melting point from top to bottom. Cold glaciers or polar glaciers are common in the Arctic and Antarctic. The extraordinary thickness of the continental glaciers in these regions results in temperature increases with depth; despite rising temperatures with depth, melting does not occur as readily within cold glaciers. However, melting at depth can occur in cold glaciers. Heat transfer via conduction between the bedrock and bottom of the glacier produces seasonal meltwater. This meltwater lubricates the bedrock and facilitates glacial movement.

## Polythermal glaciers

Although attempts have been made to simplify glaciers into two temperature categories – warm or cold, anomalies do exist. **Polythermal glaciers** exhibit both warm and cold temperature characteristics. Ice sheets, for example, vary in depth and temperature from the center of accumulation to its margins. This variability results in differing patterns of movement, erosion and landscape modification (Hambrey and Alean 2004).

# Chapter 3 Review

## Key terms

Alpine glaciers                    Highland ice fields

Continental glaciers               Piedmont glaciers

Nunatak                            Snowline

Cold glaciers                      Ice shelves

Warm glaciers                      Ice aprons

Polythermal glaciers               Ice fringes

Tidewater glaciers                 Glacierets

## Concept review and critical thinking

1. What is the snowline? How is the snowline affected by latitude?

2. Compare and contrast an alpine glacier with a continental glacier.

3. If you were a glaciologist in the field, what techniques would you use to classify a glacier by temperature?

4. Identify two characteristics of each type of glacier: cold, warm and polythermal.

## Google Earth analysis

### Instructions

Fly to the following locations and classify the glaciers as either "alpine" or "continental."

1. −85.376196, 66.947702

2. 46.852610, −121.760339

3. 73.234705, −39.161929

4. 46.502494, 8.033875

Find two alpine glaciers near the poles and two alpine glaciers close to the equator. Determine the elevation of each of their snowlines. How do their snowline elevations differ? Explain why these differences exist.

# HOW DO GLACIERS MOVE?

After a single visit to a glacier, one might be tempted to conclude that glaciers do not move. After all, how could a mass of ice sometimes exceeding three thousand meters in thickness physically transport itself from point A to point B? Repeated visits to the same glacier, however, would reveal an entirely different story. The seemingly imperceptible changes within a glacier manifest themselves on the surface over time. Surface changes such as the formation of crevasses, as well as the displacement and arrival of new rocks provide evidence of the relentless process of motion. This motion can also be detected by listening; as a glacier moves downhill the ice cracks and creaks. These sounds are products of the internal forces within the glacier.

## Mechanisms of Ice Flow

Glacier movement is controlled by gravity working in concert with the internal forces of the ice. More specifically, geologists Matthew Bennett and Neil Glasser suggest that glaciers move by three major processes: 1) **internal deformation**, 2) **basal sliding** and 3) **subglacial bed deformation** (2009, 51).

### Internal deformation

Internal deformation refers to the transformation of the internal crystalline structures of glacial ice. As discussed earlier, when snow changes to glacier ice, it deforms internally due to the overlying weight of material and the force of gravity. Such deformation of ice in response to stress occurs through the processes of **creep** and **fracture**. Of the two, creep is the dominant process driving glacial motion. Creep originates when stress overwhelms the crystalline structure of ice causing a series of dislocations between crystals. Under these conditions the ice is deformed into a viscoplastic material, which behaves like soft putty. This plastic-like material is capable of moving without fracturing. Benn and Evans compare this movement to the behavior of metals their temperatures reach melting point (2010, 115). When the ice cannot adjust fast enough to the glacier's internal stresses, a series of folds and fractures may form. Such ice structures will be discussed in greater detail in Chapter 5.

### Basal sliding

Another mechanism that contributes to glacier movement is basal sliding. Meltwater that accumulates at the base of the glacier lubricates the solid bedrock below. These lubricated films reduce friction between the glacier and consolidated rock surface allowing the ice mass to flow. According to geologists David E. Sugden and Brian S. John, basal slip can account for up to 90 percent of a glacier's movement (1976, 23).

Basal slip, however, does not produce continuous movement. The solid bedrock below the glacier is rarely flat and smooth. Consequently, glacier movement is said to have stick-slip properties, which are characterized by jerky movements. Ridges and other protrusions retard movement by influencing localized melting and freezing as well as by obstructing their flow (Menzies 1995).

### Bed deformation

A third mechanism of glacier movement is bed deformation. While basal slip is a common process of glacier movement, not all glaciers "glide" across solid bedrock. Unconsolidated sediment comprised of an unsorted mixture of clay, silt, sand, pebbles, cobbles and boulders provide a deformable surface of movement for many glaciers. When saturated with water, the resistance between sediment grains is reduced. This allows the sediments to behave in what Bennett and Glasser describe as a "slurry-like mass" (2009, 54). The shear forces of the glacier applied to the sediment, in turn propels the glacier forward as the subglacial debris deforms and reduces its frictional resistance.

## Rates of Movement

The rate of glacial movement while always relatively slow is highly variable. Slow-moving glaciers move only a few meters a year. Most glaciers according to Bennett and Glasser have velocities between 3 and 300m per year with some containing potential portions that flow 1–2km per year along steep slopes (2009, 67). Outlet glaciers and ice streams that terminate at the sea have been observed to move up to 12 kilometers in one year. This flow, however, is not always continuous. In fact many glaciers stagnate for many years and then surge over great distances unpredictably. When studying the patterns and rates of ice flow in a glacier it is important to distinguish between **longitudinal movement** and **transverse movement**.

## Longitudinal movement

Longitudinal movement or vertical movement varies in flow rates from the surface to the base of a glacier. Frictional resistance between the base of a glacier and the underlying bedrock typically slows the flow at the bottom of the glacier, resulting in faster movement at the surface. Several experiments have been performed to explain the variations in longitudinal velocity profiles. Glaciologists conduct these experiments by inserting a vertical bendable pipe into a glacier extending from the surface to its base. After several years of observations, these experiments demonstrate a consistent pattern. First, the entire pipe moves downhill from its original position. Second, the pipe bends into a curve that shows greater movement at the top than at the bottom. The rate of change, however, is highest toward the base of the glacier where internal deformation is the greatest. The top portion of the pipe appears unbent – yet it moves from its original position. Basal slipping accounts for this movement – the rate of which is dependent upon the steepness of slope and irregularities on the bedrock floor.

In an effort to maintain a stable longitudinal-velocity profile and maintain continuous flow a glacier needs to adjust to changing accumulation and wastage. As discussed earlier, accumulation occurs at the head of a glacier and wastage dominates at the glacier's snout. In this scenario it seems logical to think that a glacier would then continue to thicken at its head and thin at the snout. This pattern, however, would quickly put an end to a glacier's movement. To prevent this from happening, glaciers at equilibrium, change their inclination of flow direction (Sharp 1988). Glaciers create a counterbalance to excessive accumulation and melting by compressing and extending their flow. For example, the zone of accumulation is typically found at higher elevations with steep descents. Faster velocities of ice flow extend the glacier, thinning it out. On the other hand, the zone of ablation is characterized by lower gradients where flow velocities decrease. Compressing flow occurs in these portions of the glacier, which has a tendency to thicken the ice.

## Transverse movement

Transverse surface-velocity profiles reveal the effect of channel shape on flow rates. As pointed out earlier, ice sheet flow is unconfined by the underlying topography and velocity is typically slowed by friction at the glacier's base. Many alpine glaciers, however, are confined by valley walls. Frictional resistance along the edges can reduce flow by 80 percent (Menzies 1995, 165). Consequently, a valley glacier's velocity is greatest at the center of the flow closest to the surface. This parabolic shape associated with velocity gradients is frequently observed in valley glaciers with symmetrical U-shaped valleys. Simple experiments have been designed to observe this pattern. In fact, experiments were conducted in the Alps as early as the 1800s where boulders were laid out across a glacier in a straight line perpendicular to the slope from one valley wall to the other. After a year of observation, the line of boulders was no longer straight. The boulders at the glaciers center exhibited greater downslope displacement than the boulders near the valley's walls.

In addition to the topography and the geological environment, there are other variables that affect glacier movement. As discussed earlier, climate helps to produce both warm and cold glaciers. Rapidly flowing glaciers are more common among warm glaciers where the ice is at the pressure-melting temperature. Under these conditions, percolating meltwater enhances creep and basal slipping (Sugden and John 1976). On the other hand, cold glaciers found in polar environments typically exhibit slower flow rates.

# Chapter 4 Review

## Key terms

Basal sliding                           Longitudinal movement

Internal deformation                    Creep

Bed deformation                         Fracture

Transverse movement

## Concept review and critical thinking

1. Identify and explain the three internal processes responsible for a glacier's movement.

2. Using 20 large boulders, explain how you could set up an experiment to test the differences in transverse movements of a glacier. Explain the expected results.

3. Using a long bendable pipe, explain how you could set up an experiment to test the differences in longitudinal movements of a glacier. Explain the expected results.

## Google Earth analysis

The Aletsch Glacier is the largest glacier in the Swiss Alps. The alpine glacier spans a length of approximately twenty-three kilometers and covers an area of 120km². Just south of the main summits of the Bernese Alps, the glacier emerges as three smaller valley glaciers join together in a vast flat area of snow and ice known as Concordia. At this junction the glaciers thickness is estimated to be over a kilometer in depth. After the confluence, the main stem of the glacier flows south and then southwest into the Rhone Valley.

## Instructions

Fly to the Aletsch Glacier in the Swiss Alps using Google Earth. Type in the latitude/longitude coordinates found below into the "Fly To" box.

*Aletsch Glacier: 46.469190, 8.071278.*

Zoom into the location just close enough so that you can see the entire glacier – from where it originates in the north to where it flows southwest. Using the print-screen function on your keyboard, print the image located using Google Earth.

Using a red-color pencil, draw an arrow on the main stem of the Aletsch Glacier to indicate where the fastest flowing ice can be found according to the principles of transverse movements. Using a blue-color pencil, draw an arrow to indicate where the slowest ice velocities can be found. Next, label the location on the glacier where you would expect to find compressing ice flow and where you would expect to find extending ice flow.

# ICE STRUCTURES

Anatomically speaking, glaciers have a head or starting point in the area of accumulation, a tongue that extends to the lower reaches of the glacier, and a snout or terminus in the area of ablation. While useful on a large scale, these macrofeatures fail to recognize the structural complexities found on top of and within the ice. Surface and subsurface ice forms are controlled by a glacier's movement as well as by the processes of accumulation and ablation.

## Surface Ice Structures

### Ogives

The surface of a glacier is far from a featureless white blanket. A glacier's movement produces a variety of dramatic surface-ice structures. **Ogives**, for example are repetitive bands or waves that originate at the base of icefalls where blocks of ice collapse. Ogives exhibit a convex pattern that curve downhill. The curvature, according to Benn and Evans is a reflection of the greater velocity of the central part of the glacier compared to its flanks (2010, 140). While these patterns are consistent, glaciologists have recognized two types of ogives. **Band ogives** consist of alternating bands of light and dark ice. Individual bands are typically several meters wide. Many glaciologists have asserted that pairs of light and dark bands are annual features. In other words, the light and dark nature of these bands might be attributed to the season in which the ice collapsed. The white bands, rich in air bubbles, are thought to have formed in winter, a time of intense compression. The dark bands were probably formed during summer where debris discolored the melting ice giving it a dirty appearance. Similarly **wave ogives** demonstrate a similar pattern, but with alternating troughs and ridges. Troughs form during the summer months of ablation and the ridges form the excess snow during winter. Thus, by counting the number of ogives it is possible to determine how long it has taken a glacier to move from an icefall to its snout (Sharp 1988).

In addition to the annual accounting provided by ogives, the distance between bands also provide an accurate indicator of a glacier's speed. Generally, the farther the bands are spaced from each other the faster the glacier's flow. Where the glacier's velocity decreases downslope, the ogives become more closely spaced as the glacial flow compresses.

## Crevasses

Crevasses are quite possibly the greatest danger of glacier travel. These V-shaped cracks vary from a few millimeters to several meters in width. Their danger resides

**Figure 11.** Measuring snowpack in a crevasse on the Easton Glacier, Mount Baker

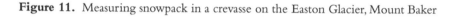

**Figure 12.** The Khumbu Icefall, Mount Everest

Source: Uwe Gille.

not so much in their width, but rather in their depth. Many of these vertical blue chasms are many times deeper than they are wide. Adding to the already-present danger is the fact that crevasses are not always visible at the surface as drifting snow frequently covers them. Many unknowing glacier travelers have fallen to their death while crossing over these unsound snow bridges.

These perilous journeys have led many to speculate that some crevasses are "bottomless." Although this makes for a great story, it is simply not true. The depth of a crevasse is inherently limited by the nature of ice under pressure. Close to the surface the ice behaves in a brittle manner. This rigidity of the ice promotes buckling as it fractures from stress applied by a glacier's movement. These cracks at the surface usually only extend to a depth of roughly 25–36m. Geologists Michael Hambrey and Jürg Alean confirm that at these depths the brittle nature of the ice is overcome by internal deformation (2004, 83). Instead, the plastic behavior of ice under pressure promotes creep and prevents additional fracturing.

The geometry of crevasses is determined by the underlying topography, as well as the direction and rate of ice flow (Sharp 1988). Steep slopes with accelerating flow rates, for example, will often experience intense crevassing. Initially, these crevasses form perpendicular to the direction of flow. This apparent organization,

**Figure 13.** Crevasses and icefalls as seen from space on an unnamed glacier in the Himalayan mountain range

Source: NASA.

however, turns to chaos as the crevasses break up into icefalls. **Seracs**, or massive towers of unstable ice are commonly found in icefalls. Many climbers have fallen victim to the collapse of massive seracs in the infamous Khumbu Icefall on Mount Everest (Figure 12).

# Subsurface Ice Structures

## *Layers of accumulation*

Despite the obvious dangers crevasses pose, glaciologists seem attracted to them. This is not only a testament to their sense of adventure, but more importantly to their quest for knowledge. Crevasses serve as a portal to the subsurface structures and complex processes within a glacier. Almost immediately after peering down into a crevasse, one obvious subsurface structure becomes apparent – layers. Like sedimentary rocks, glaciers develop a distinct stratification from repeated depositions. Unlike the layers of sediments in sedimentary rocks, glacier stratification is a result of seasonal deposition of snowfall in the accumulation area. Each layer represents one year of accumulation. These layers are often several meters thick and show a gradation of color and texture from the bottom to the top. The bottom of each layer is typically comprised of coarse-grained blue ice that accumulated and deformed under the weight of overlying snow during the winter months. A thin dark blue band often separates this portion of a layer from the ice that formed above the following summer. The finer-grained white summer ice appears different due to its frequent exposure to freeze–thaw periods (Benn and Evans 2010).

As a result of this sedimentary stratification, principles of relative and absolute dating can be applied to the geologic history of a glacier. For example, the **law of superposition** states that the oldest layers are found on the bottom and the most recently formed layers are closest to the top. This works like a messy desk drawer at home. When shifting through papers in search of a document several years old, one would have to do some digging through recent papers at the top of the pile to find the older ones at the bottom. Knowing that each layer represents one year of accumulation, one might be able to determine the age of a glacier by simply counting the layers as one would do while counting the rings of a tree.

In doing this, however, one might encounter a few problems. Some layers, for example, may be missing. Layers exposed at the zone of ablation are often lost to wastage. These truncated layers then can be covered over by additional accumulation. The boundaries between lost layers and new accumulation are known as **unconformities**; they represent missing puzzle pieces that may account for a significant loss in the geologic record.

## Foliation, folding and faulting

Further adding to the complications of determining the age of a glacier by counting layers is the fact that the strata within glacial ice are often folded or faulted and therefore discontinuous. The conditions and processes associated with ice under pressure deform the continuous layered structures within the glacier. It is helpful to think of this in terms of the rock cycle as it relates to the metamorphism of sedimentary rock. As stated earlier, sedimentary rocks are composed of compacted and cemented sediments deposited in horizontal layers. These horizontal layers suggest stable conditions, as internal forces have not influenced them, (tectonics). When these layers are out of the original horizontal position (folded, faulted, or overturned), however, it becomes evident that tectonic forces are operating in the landscape. Rocks in these deformed landscapes are often transformed from sedimentary in origin to metamorphic rocks. The intense heat and pressure associated with colliding lithospheric plates, reconstitute the mineral makeup of the rock through recrystallization and the realignment of minerals. The realignment of minerals results in **foliation** where various minerals band together in response to compressional forces. Under such pressure, rocks behave in a plastic manner and are capable of additional plastic flow and folding.

Like the transformation of sedimentary rock to metamorphic rock in the rock cycle, the stratification of glacial ice becomes highly deformed under pressure. In fact, both folding and foliation occur at depth in a glacier under plastic flow conditions. The orderly and continuous stratification of the subsurface ice forms become thin, highly deformed and discontinuous in response to pressure (Sharp 1988). Foliation is commonly found in high-stress environments along glacial margins. For example, foliation is often observed along the lateral edges of valley glaciers where the ice comes into contact with the rigid valley walls. Likewise, when two valley glaciers converge or an ice stream from a tributary merges with an ice stream of a different velocity; stress builds up causing the ice to recrystallize. The result is ice that is transformed in both crystal size and air-bubble content. The foliated ice structures align themselves perpendicular to the forces of compression and parallel to the direction of flow. This alignment often manifests itself at the surface of a glacier in the form of a furrowed surface. Thus, foliation has been used as navigational tool for misguided glacier travelers caught in whiteout conditions.

**Folding** often accompanies foliation. Plastic flow allows the once rigid ice to bend into dramatic shapes. Occasionally changing conditions in the ice overwhelms the plasticity. Polythermal glaciers, for example, alternate between conditions at or below the pressure melting point. If basal sliding encounters dramatic freeze–thaw conditions, the once pliable ice at the glacier's base may fracture. These cracks produce vertical or lateral faults. Low-angled fractures originating at a glacier's base develop **thrust faults** where the ice upstream rides up over the ice downstream. Thrust faults are common among glaciers that experience periodic surges in which ice moves 10–100 times the velocity of normal movement. In addition to increased velocities, the transport of ice and debris often increases at these interfaces (Hambrey and Alean 2004).

# Chapter 5 Review

## Key terms

Ogives                                    Unconformities

Band ogives                               Foliation

Wave ogives                               Folding

Crevasses                                 Thrust fault

Law of superposition                      Seracs

## Concept review and critical thinking

1. What factors and processes influence the formation of both surface and subsurface ice structures?

2. Compare and contrast band ogives with wave ogives.

3. Explain how subsurface ice structures can be used to determine a glacier's relative age. Identify two factors that may limit this dating method.

4. How do unconformities develop within a glacier?

5. Distinguish between foliation, folding and faulting within a glacier.

## Google Earth analysis

The Gilkey Glacier is located in the Juneau Icefield just north of Juneau, Alaska. Several tributary glaciers join the 32km Gilkey Glacier before draining into a lake at the Gilkey terminus. Just south of the Gilkey trench is a vast icefall. Stemming from the icefall is a series of parabolic wave ogives that can easily be identified by a series of crests and troughs that point down glacier. The ogives form annually and therefore provide a means to evaluate glacial velocity. By counting the number of bands and measuring the distance from the icefall to the glacier's terminus, you can determine a rate of movement.

## Instructions

Take a trip the, Gilkey Glacier, Juneau Icefield using Google Earth. Type in the latitude/longitude coordinates found below into the "Fly To" box.

*Gilkey Glacier: 58.829001, −134.292084*

1. Examine the train of gives (parabolic bands of light and dark colors) that extend to the southwest from the icefall at the center of your map. Note: the icefall itself is not visible with the current resolution.

2. Count the number of ogives from crest to crest. Each one represents one year.

3. Use the "Ruler Tool" to determine the distance in meters from the icefall to where the ogives stop.

4. Calculate the glacier's velocity.

$$\text{Glacier's velocity} = \frac{\text{Distance}}{\text{Years}}$$

5. What does the parabolic shape of the ogives indicate about the transverse movement of the glacier?

# 6

# GLACIAL EROSION

Erosion is the unrelenting process of moving material at the Earth's surface. In fact, five natural agents of erosion – gravity, wind, running water, wave action and glaciers, are constantly reshaping the surface of the Earth. Gravity is first on the list because without it, the four other agents of erosion would not work. Rivers and glaciers, for example, transport sediment under the influence of gravity by flowing downhill. While gravity is the primary force behind all agents of erosion, it can operate independently. Mass movements, such as landslides, mudflows and avalanches are all events capable of commanding our attention. Yet the power of gravitational erosion often lies in the less dramatic events. Most gravitational erosion occurs painstakingly slow in the form of a slump or creep where sediments move imperceptibly from areas of high elevation to areas of low elevation. This is where glaciers pick up the trail as they continue to transport these eroded particles to even lower elevations. Many scientists refer to glaciers as nature's conveyor belt, as few natural processes can rival their ability to move large amounts of debris such great distances. Before delving into the basic processes of glacial erosion, let's first take a look at the types of glacial debris and transport. Glacial debris enters glacial systems from two sources. Bennett and Glasser assert that debris entrainment occurs at the surface of a glacier (**supraglacial debris**) or at the glacier bed (**subglacial debris**) (2009, 185).

## Glacial Debris Entrainment

### Supraglacial sources

The rocks and sediments that collect on top of a glacier from mass movements are known as supraglacial debris; the major source for this accumulated sediment is rock fall. Glaciers have a tendency to oversteepen and destabilize valley walls.

These rocks at the glacier's flanks are further undermined by frost action. Water seeping into the cracks of rocks alternates between periods of freezing and thawing, which results in periodic expansion and contraction. Over time, the rocks split away from the rock face falling into the valley below. The majority of this rock fall accumulates on top of the glacier at the base of the valley wall where it is set to move again on the glacier downhill.

Another important source of supraglacial debris is snow and ice avalanching. In high alpine environments such as the European Alps and the Himalayas, large quantities of snow can accumulate precariously on steep slopes. Repeated avalanches in these regions have been reported to provide the single greatest source of supraglacial debris. Such destructive avalanches scour and pluck the underlying rock surface transporting a mixture of ice, snow and rock, which accumulate on the surfaces of valley glaciers below (Benn and Evans 2010).

### Subglacial sources

Not all rocks remain on the glacier's surface. Some of the supraglacial debris becomes incorporated into the interior of a glacier as snow blankets the accumulated sediments or sediments fall into crevasses. Such **englacial debris**, however, does not only become entrained through the surface of a glacier. Most ice sheets and ice caps, for example, are devoid of supraglacial debris. These continental glaciers typically submerge mountains and rarely have rocks exposed at their flanks. Yet continental glaciers still contain an enormous amount of entrained debris. Thus, it can be inferred that a great deal of glacial erosion occurs along the bottom of a glacier.

# Erosional Processes

Understanding how this subglacial erosion takes place is difficult. Where does the entrained debris come from? Getting to the bed of an active glacier is not easy and has only been accomplished by a select few. Techniques include peering into crevasses, using tunnels dug out by mining companies, as well as remote-sensing technologies. Although these methods have revealed a great deal of information about glacial erosion, most of our understanding is derived from the polished bedrock and sediment features left in the wake of a glacier's retreat. These postglacial landscapes help to explain the often-inscrutable processes of glacial erosion. From these investigations, the basic processes of glacial erosion can be broken down into three major mechanisms: **abrasion**, **plucking** and the action of **basal meltwater**.

## Abrasion

Abrasion refers to the process where basal debris in the ice, scours across the underlying bedrock. Evidence of this can be accounted for in the small-scale features of the bedrock. **Glacial striations** or scratches are etched into the surface of rock that was overridden by sediment-laced ice. These scratches are typically no deeper than a few millimeters but have been measured over several meters long. Striations are oriented parallel to local ice movement and therefore have allowed scientists to reconstruct local ice-flow patterns. Bedrock containing striations often appears smooth and polished. Hard, angular minerals such as quartz and some silicates act like sandpaper on the bedrock below while finer particles of silt commonly referred to as **glacial flour** polish the bedrock surface (Sharp 1988). Larger rocks and cobbles embedded at the base of a glacier, judder over the rock, producing small fracture marks or cracks in the bedrock known as chattermarks and crescentic gouges. In doing this, the basal debris becomes more rounded as it grinds over rock through the process of erosion.

## Plucking

A second method of basal glacial erosion is plucking. Also known as quarrying, plucking is the process in which basal ice fractures the bed and dislodges rock fragments. Glaciers are particularly effective at dislodging blocks from weakened bedrock. Pre-existing weakened bedrocks often get their start in a landscape prior to glaciation. In these harsh environments, the bedrock is exposed to the atmosphere where it is chemically and physically weathered. The bedrock is particularly weakened by frost action – a physical weathering process that results from the expansion and contraction of freezing and melting water. This weathering process fractures the bedrock and creates lines of weakness that glaciers can exploit.

Additional erosion in these preglacial environments can enhance rock weakening through pressure unloading. As the weight of overlying rock layers is removed from the surface, the bedrock expands and cracks parallel to the surface when pressure is reduced. These cracks are known as dilatation joints and are often exploited by glacial erosion.

Not all block weakening occurs prior to glaciation. There is evidence that active glaciers can fracture and weaken the underlying bedrock by their own doing. For example, both frost action and dilatation joints occur beneath active glacial ice. Glacial meltwater located at the base of a glacier manages to penetrate cracks in the bedrock. Freeze–thaw cycles in a polythermal glacier can transform a small crack into a significant fracture.

**Figure 14.** Diagram of glacial plucking and abrasion

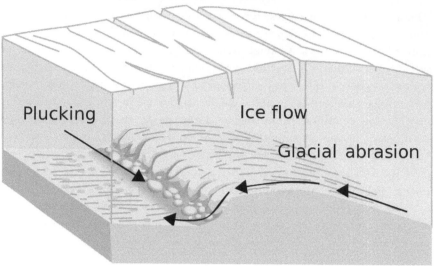

Source: Luis Maria Benitez.

Such subglacial weathering, however, cannot account for the dramatic plucking of deep valley glaciers. Geologists David Sugden and Brian John suggest that quarrying at this level might be best rationalized by large-scale dilatation jointing (1976, 157). Dilatation jointing at this scale occurs where a glacier creates a large depression in the bedrock. At first it would seem that the weight of the ice would counteract any expansion of the underlying bedrock. Over time, however, as the depression deepens, the weight of the ice becomes insignificant compared to the lost pressure exerted by the rock that once covered the area. Consequently, the rocks at depth will rebound and produce fractures in parallel sheets due to a release in pressure. Glaciers would then be free to excavate and pluck these weakened rock faces.

### Basal meltwater

As previously mentioned, block weakening provides glaciers with the opportunity for rock dislodgement. Plucking rocks from fractured bedrock is then subsequently achieved through effective glacial pressure and friction. Menzies suggests that a major factor controlling these two variables is the fluctuation of basal meltwater (1995, 256). Water discharge at the base of the glacier varies daily; these diurnal fluctuations create different conditions in subglacial cavities. During periods of high water pressure, the ice is often separated from the bedrock below. This high ice-bed separation reduces the frictional drag along the bedrock. During periods of low water pressure, the weight of the glacier that was supported by the pressurized meltwater returns to the bed where it slides with significant force. The fluctuation in ice-bed separation associated with pressurized meltwater is a major factor in the effective pressure and friction required for rock dislodgement (Menzies 1995). Once dislodged from the bed, freezing onto the base of the glacier incorporates the basal debris. Tightly embedded into the ice, the rocks become powerful abraders and pluckers that further erode the bedrock.

# Chapter 6 Review

## Key terms

Supraglacial debris                 Plucking

Englacial debris                    Basal meltwater

Subglacial debris                   Glacial flour

Abrasion                            Glacial striations

## Concept review and critical thinking

1. Define erosion and identify Earth's five natural agents of erosion.

2. Explain how supraglacial debris accumulates on a glacier. How does supraglacial debris become entrained within a glacier?

3. Why are ice sheets typically devoid of supraglacial debris?

4. How do scientists study subglacial erosion?

5. Identify and explain the three major mechanisms of glacial erosion.

# LANDFORMS OF GLACIAL EROSION

While the small-scale features of glaciated bedrock are useful in understanding the mechanisms of glacial erosion, perhaps the most impressive products of glacial erosion are the intermediate and large-scale landforms. Let's redirect our attention away from the details of the rock surface and examine the topographic bedrock manifestations of glacial erosion.

## Intermediate-Scale Features of Glacial Erosion

### Roches moutonnées

As pointed out in Chapter 6, abrasion is the primary force responsible for producing the small-scale rock-surface features such as glacial striations and grooves. In the formation of intermediate and large landforms, however, plucking is generally the process that dominates. **Roches moutonnées**, notwithstanding, are products of both processes. Named after their resemblance to a French wig popular during the eighteenth century, roches moutonnées are asymmetrical hills commonly found at the bottom of deglaciated valleys. The pronounced asymmetry of these hills are characterized by one side smoothed and molded by abrasion and the other side sharply steepened and truncated by plucking. Sharp suggests that the aspects of each of these distinct rock faces speak to the direction of ice flow (1988, 101). The rounded, striated, polished flank subjected to intense abrasion faces upstream of the glacier's advance. The steep, angular, craggy flank subjected to plucking is found on the leeside. The plucked face is not only steeper but it is often higher as a result of plucking on the leeside of the knoll. These prominent features in the landscape range in height from a few meters to several hundred meters. Their asymmetrical shapes are unmistakable and are often thought of as a quintessential intermediate-scale feature of glacier erosion.

**Figure 15.** A roche moutonnée in the Cadair Idris Valley, Snowdonia, Wales

## *Whalebacks*

Whalebacks are intermediate-scale landforms produced entirely from glacial abrasion. These symmetrical hills are bedrock protrusions that have been smoothed and rounded on all sides by a glacier. Named after their resemblance to the back of a whale breaching the ocean surface, whalebacks lack the plucked lee faces found in roches moutonnées. Benn and Evans argue that their distinctive symmetry is due to high-average ice-overburden pressures that are conducive to glacial abrasion and the suppression of plucking (2010, 280). Consequently, many whalebacks appear smooth and polished and contain striations that run their entire length. Whalebacks typically measure between 1 to 2m in height and 1.5 to 3m in length.

# Large-Scale Features of Glacial Erosion

## *Glacial troughs and fjords*

Troughs or glaciated valleys are perhaps the most well-known landforms connected to glacial erosion. These often-symmetrical alpine valleys are commonly described in cross section as U-shaped, but this description is a

**Figure 16.** Glacial trough or U–shaped valley in Mount Baker Snoqualmie National Forest

Source: Walter Siegmund.

bit overstated. Instead, glaciologists favor the term parabolic since the valley wall's steepness is rarely vertical. More often than not, these troughs manifest prior to the glacial advancement. In fact, an entirely different agent of erosion instigates this process. Running water or stream erosion initiates the valley's formation by carving through the bedrock. These rivers rarely follow a straight line and instead follow the path of least resistance. The results are winding, deep, V-shaped canyons with narrow floors and steep valley walls. Unfortunately for glaciers, these meandering, narrow river canyons are not conducive for ice flow. Therefore, when a glacier moves down through the river channels it immediately begins to modify the valley. These modifications include widening, straightening and deepening the original river valley. The results are staggering. The once steep, wandering river valley is transformed into a gently sloping, straightened parabolic valley. The troughs are deeper and broader and are capable of discharging vast quantities of ice. In the wake of a receding glacier, a variety of fascinating features can be found in and around the newly formed valley.

**Figure 17.** Glacial trough or U-shaped valley in the Swiss Alps

**Fjords** are long narrow inlets bound on both sides by steep cliffs. Bennett and Glasser define fjords as "drowned glacial troughs" (2009, 169). Confined by their topography, valley glaciers have a tendency to overdeepen their floors – sometimes to depths well below sea level. Fjords form where these valley glaciers extend into the sea. Like glacial troughs, fjords have parabolic cross sections and are commonly found in the Arctic and Antarctic. Other well-developed fjord coastlines can be found in South America, Norway, southern Alaska and New Zealand.

## Hanging valleys

Other incredible sites found within glacial troughs and fjords are **hanging valleys**. Prior to glacial advancement, the main stem of a river carves out a trunk valley. This trunk drains all the tributaries that flow into it at accordant junctions; these mergers occur at a common elevation. As pointed out earlier, when a glacier moves down through the trunk valley it not only widens it, but it deepens it as well. Glacial erosion may deepen these valleys hundreds of meters below their original elevation. The result is a discordant junction, where the tributaries are perched high above the newly carved glaciated valley. Spectacular waterfalls are often observed at these glacially formed discordant junctions.

**Figure 18.** Geiranger Fjord, Norway

Source: Frederic de Goldschmidt.

## Cirques, arêtes and horns

Glacial erosion leaves several tell-tale signs at the heads of mountain glaciers. **Cirques**, **arêtes** and **horns** provide insightful evidence of erosion at higher elevations. A cirque is a hollowed out bowl-like basin formed at the head of a glacier. These excavated areas typically have steep headwalls and sidewalls. These walls bind cirque glaciers where erosion extends downwards. However, overtime the glacier can turn its erosive power 180 degrees and begin devouring the mountain headward (Sharp 1988). This occurs where thin ice accumulates at the head and sides. Freeze–thaw cycles break up the rock, and the glacier undercuts the head wall. This sapping process steepens the head wall and enlarges the glacier. The supraglacial debris is moved downslope along the edge

**Figure 19.** Fjords found along the Southern Patagonian Ice Field

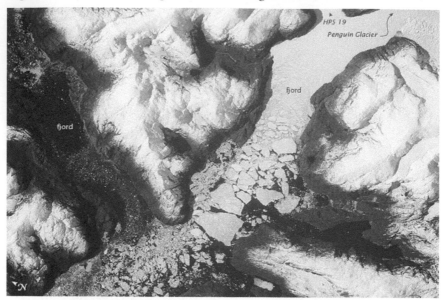

Source: NASA.

**Figure 20.** The Garden Wall, an arête in Glacier National Park

**Figure 21.** Glacial tarn formation

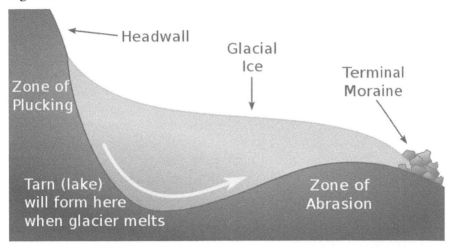

**Figure 22.** The Matterhorn

Source: Marcel Wiesweg.

creating an embankment that serves to further shape the semicircular basin. Over time with additional excavation, the overdeepened bowl-like basin may produce a small lake or **tarn** (Figure 21).

Frequently, glaciers team up to launch an all-out attack on a mountain. When three or four facets of a mountain are subjected to the sapping processes that form cirques, a horn may develop. The Matterhorn in the Swiss Alps, with its pyramidal spire culminating at a sharp point, serves as a stunning example of such glacial erosion (Figure 22). These features may be connected together by razor-edge ridges or arêtes that are bound on both sides by steep straight slopes (Figure 20). These topographic divides produced by two opposing cirques have provided mountaineers with challenging routes to the some of the world's most famous peaks.

# Chapter 7 Review

## Key terms

| | |
|---|---|
| Troughs | Roches moutonnées |
| Cirques | Hanging valleys |
| Arêtes | Fjords |
| Horns | Tarns |

## Concept review and critical thinking

1. What erosional processes dominate in the formation of intermediate and large-scale landscape features?

2. Compare and contrast roches moutonnées with whalebacks.

3. Explain how glacial troughs and fjords form.

4. Why are hanging valleys perched higher than the main valley's trough?

5. Distinguish between cirques, arêtes and horns.

### *Google Earth analysis*

### *Instructions*

Fly to the following coordinates and identify the erosional landform depicted. Each of the erosional landforms listed in the key terms above will be used once. Many recognizable landforms may appear at each location, but the landforms to be identified will appear at the center of the viewing screen.

1. 45.976392, 7.659258

2. 52.693792, −3.895991 (zoom in to the small rock formation)

3. 48.828584, −113.744927

4. 44.261710, −71.296451

5. 37.716719, −119.646523

6. 46.423224, 7.829054

7. 52.703824, −3.913891

8. 62.112513, 6.942071

# 8

# GLACIAL DEPOSITION

Like a crime-scene investigator, a geologist reconstructs the geologic past with clues left behind in the landscape. Glaciologists rarely have a hard time finding clues because glaciers behave like sloppy criminals. In the wake of glacial recession, unique deposits and peculiar landscape features are left behind and often stick out like sore thumbs. Glaciers seem to have the intention of "stealing" a lot of rock material by simply plucking it from the landscape. Through erosion, they may even transport it great distances. Eventually, however, they inevitably run out of steam and drop the entrained material; this process of dropping the transported material is known as deposition. There are two major categories of glacial deposition of sediment on land – direct and indirect deposition.

## Types of Glacial Deposition

### Direct glacial deposition

Direct glacial sedimentation occurs when material is laid down from the ice itself. These debris deposits are commonly known as **glacial till** and are characterized by their untidy nature. As previously mentioned, glaciers are notoriously messy. Till is a jumbled mixture of sediments ranging in size from grains of clay to large boulders; the only pattern to these deposits is the lack of pattern. The sediments are not arranged in layers and are poorly sorted. As a general rule the majority of the sediment matrix is a fine-grained mixture of clay and silt, yet from location to location till is highly variable. The stones embedded in this disarrayed matrix can be large or small, rounded or angular; some deposits are striated, some are not.

This high variability in glacial till is in part a function of the different environments of deposition. In response to these different depositional conditions, there are several mechanisms of direct deposition and thus types of till.

**Figure 23.** Angular glacial erratic on Lembert Dome in Yosemite National Park

Source: Daniel Mayer.

For example, **lodgement till** is deposited when the underlying bed snares a rock entrained within a glacier. When the resistance of the underlying bed exceeds the momentum of the overlying ice movement, the rock stops moving. Another process of direct deposition is the direct release of material from basal melting of stagnant or slow-moving sediment-rich ice. This sedimentation produces subglacial **melt-out till**. The size of the entrained material released from this process is dependent upon the degree of melting. Generally, the higher the degree of melting, the larger the material deposited.

Many deposits of direct deposition are foreign to the underlying bedrock they are found on. This clue provides evidence that the rocks and sediments may have been transported great distances from their place of origin – perhaps hundreds to thousands of kilometers. Large boulders known as **glacial erratics** appear to be randomly flung across many postglacial landscapes. Robert Sharp points out that some of these erratics turn out to be critical pieces of evidence. Known as indicators, these large boulders may contain minerals and other source material that is traceable to its point of origin. Many portions of the upper midwestern United States, for example, contain countless glacial erratics. The compositions of these large boulders are made up of intergrown crystals of several different silicate minerals. Geologists have determined that these rocks are entirely different from the underlying bedrock terrain. Following the boulder train north to the point of origin, geologists determined the source region for many of these erratics to be Calgary, Canada (Sharp 1988).

## Indirect glacial deposition

Unlike direct glacial deposits, indirect glacial sedimentation produces deposits that are not restricted to the general vicinity of the glacier. Indirect deposition may occur within a glacier's boundary or extend well beyond its margins; streams of glacial meltwater typically transport these sediments. The sediment that is transported twice over by glaciers and then running water is known as **glacifluvial material**. The physical characteristics of glacifluvial materials differ dramatically in appearance and organization from glacial till. As discussed earlier, till is deposited without rhyme or reason; the sediment is deposited all at once in an unsorted heap. Flowing meltwater, however, tends to be a bit more organized as a depositional system.

This organization is in part a function of the changing velocities of the transporting medium. As a meltwater stream exits a glacier it carries a sediment load with a wide array of particle sizes. The microscopic particles are dissolved and carried in solution. Clay, silt and sand give the water a milky appearance as they are carried in the water column via suspension. The larger particles such as pebbles, cobbles and boulders are transported by saltation or tumbling along the streambed. Faster-moving streams can transport larger sediment and more of it. The velocity of a meltwater stream is determined by the steepness of slope and volume of water surging through the stream channel. If the slope decreases and/or the discharge is reduced, the stream will inevitably slow down. When the stream velocity decreases, the carrying capacity of the stream decreases as well. Thus, as a stream slows, deposition increases. This deposition is orderly and systematic. First to fall out of the meltwater streams are the largest sediments. Boulders, cobbles and pebbles are the heaviest items in the sediment load and are therefore the first to be deposited. Next, sand, silt and finally clay are released from the transporting system. In this way a horizontal sorting of sediment develops where the finer material is carried farther away from the initial site of deposition.

This horizontal sorting extends outward from the glacier's margins in a formation known to geologists as an **outwash plain**. Due to the high stream loads, increased discharge and steep gradient, a **braided–stream** pattern develops. According to Bennett and Glasser, this pattern emerges when a series of longitudinal bars are deposited often in the middle of the stream, parallel to the current flow (2009, 241). These bars are small at first but grow in size as fine sediment accumulates on the downstream side. Again, in accordance with horizontal sorting, grain size decreases downstream.

# Chapter 8 Review

## Key terms

Glacial till

Lodgement till

Melt-out till

Glacial erratics

Glacifluvial material

Outwash plain

Braided stream

## Concept review and critical thinking

1. What is deposition? Distinguish between direct and indirect deposition.

2. Define glacial till. Compare and contrast lodgement till with melt-out till.

3. Explain how glacial erratic can be used to determine a glacier's source region.

4. How does glacifluvial material differ from glacial till?

5. Describe the formation of a braided-stream pattern and explain how horizontal sorting contributes to the patterns observed in outwash plains.

# LANDFORMS OF GLACIAL DEPOSITION

Glacially deposited landforms are highly variable. This variability is a reflection of the wide array of depositional environments. To simplify these complexities, depositional landforms can be classified into two main groups. These groups, according to Bennett and Glasser, include landforms that are formed in an ice-marginal position and those that form in a subglacial position (2009, 247). This organizational system of course has its limitations, as there are many examples of depositional landforms that form in more than one depositional environment.

## Landforms Created in an Ice-Marginal Position

A major hotbed of landforms of glacial deposition occurs in an ice-marginal position, along the edges of present and former glaciers. Ice-marginal **moraines** for example, are large ridges of sediment debris deposited or deformed at the edge of an active glacier. These ridges vary in composition. Some moraines are composed entirely of glacial till, while others are composed of only glacifluvial deposits. Many others are a mixture of both deposit types. In spite of these variations, moraines are perhaps the most recognizable landforms of glacial deposition.

### Terminal and recessional moraines

Geologists have identified a variety of moraine types commonly associated with both valley and continental glaciers. For example, a moraine found at the farthest extent of a glacier's advance is known as a **terminal moraine**. These ridgelines run parallel to the snout of the glacier and are often deposited when the glacier halts and dumps its debris via gravity and running water. The longer the glacier stays in place, the larger the end moraine becomes. Although less dramatic in topographic relief, the terminal moraines of continental glaciers may extend hundreds of kilometers in distance and rise hundreds of meters in height.

**Figure 24.** Glacial moraines in Alberta, Canada

Source: Mark A. Wilson.

The ice sheets that covered much of North America during the last ice age, for instance, produced massive terminal moraines at their farthest southern extent. Several of these end moraines form the backbone of both Cape Cod and Long Island in the northeastern United States.

As a glacier recedes from its terminal position, a series of recessional moraines may develop. These ridges are deposited when the glacier's retreat is interrupted by periods of pause and stabilization. These upslope ridges tend to run parallel to the terminal moraine, and are sometimes larger.

## Lateral and medial moraines

Extending along both sides of a glacier's tongue are lateral moraines. Perpendicular to terminal and recessional moraines, these ridges collect a significant amount of debris along the valley walls. The debris not only comes from the entrained material within the glacier, but also from above. As the glacier scours through the landscape, the valley walls above may become oversteepened and ultimately undermined. Rock debris in the form of mass movements may add to the extent of lateral moraines, which may attain heights of over three hundred

**Figure 25.** A medial moriaine in the Upsala Glacier found in the Southern Patagonian Ice Field

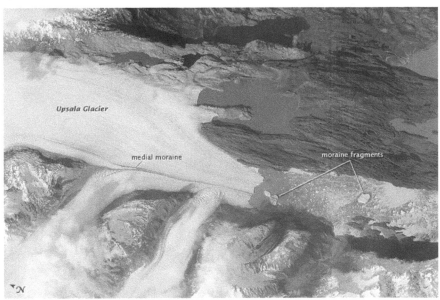

Source: NASA.

meters (Sharp 1988). At the confluence of two adjoining valley glaciers, the two interior lateral moraines of the independent glaciers fuse together into a medial moraine that runs up the center of the newly merged glacier.

# Landforms Created in a Subglacial Position

Depositional features formed in a subglacial position are an entirely different animal from those formed in the ice-marginal position. One major difference can be found in the shapes of the depositional features; subglacial-position deposits often appear to be molded.

### Drumlins

**Drumlins**, for example, are asymmetrically streamlined hills, 5–50m high and up to five hundred meters wide, produced by the advancement of a glacier moving over a previously deposited moraine. The overriding ice steepens the upstream side, leaving a gentle slope on the leeside of the spoon-shaped hill. This pattern provides geologists with clues that indicate the glacier's direction and velocity of flow. Often clumped together in swarms, drumlin fields are

**Figure 26.** Drumlin in Clew Bay, Ireland

Source: Brendan Conway.

primarily the legacy of extinct continental ice sheets and can be found in Upstate New York, New England and the Midwest prairies of North America.

### Eskers

**Eskers** are also formed in the subglacial position of ice sheets. These long winding ridges typically run parallel to the glacier's movement. Their formation is thought to occur during the end of a glacier's advance, commonly referred to as a period of deglaciation. As the thin, slow-moving ice recedes, meltwater falls to the base of the glacier through crevasses, forming subglacial streams encapsulated by walls of ice. These pressurized stream channels have the capacity to flow uphill and transport copious amounts of debris. Deposits of glacifluvial sand and gravel eventually fill in the channel. As the ice recedes, solitary and braided ridges remain – ranging in height from 20 to 200m.

## Ice Contact Features

Adding another layer of complexity to the classification of landforms of glacial deposition are ice-contact features. These features are often controlled by the interaction of glacial meltwater with an ice mass. Glacifluvial deposits often produce kame and kettle topography – or one of mounds and pits.

### Kames

Kames are conical-shaped hills deposited by superglacial meltwater. Meltwater flowing over the glacier's surface can carry a significant load of glacifluvial material. To form a kame, this meltwater pours into a hole in the ice extending to the bed of the glacier. Overtime the hole fills itself with the glacifluvial material, which is shaped by the walls of ice. Like eskers, kames are created during deglaciation when ice flow is minimal.

### Kettles

If kames represent the mounds, then the kettles exemplify the pits. Kettle holes are enclosed depressions created from the melt out of buried ice. These water-filled depressions are typically associated with glacifluvial deposits. Ice blocks detached from a glacier are often encased in moraines or outwash deposits. The size and depth of burial of the ice block according to Robert Sharp determines the characteristics of the kettle hole (1988). Depending on these factors, the dimensions of a kettle hole can range from a few meters in diameter to several kilometers, and depths can range from 1 to 50m. Ice blocks buried completely tend to produce smooth shorelines with gentle banks while ice blocks buried close to the surface or incompletely yield a kettle hole with steep banks and irregular shorelines.

## Chapter 9 Review

### Key terms

| | |
|---|---|
| Moraine | Drumlins |
| Terminal moraine | Eskers |
| Lateral moraine | Kettles |
| Medial moraine | Kames |

### Concept review and critical thinking

1. How do landforms created in the ice-marginal position differ from those that are formed in a subglacial position?

2. What is a moraine? Distinguish between a terminal, recessional, lateral and a medial moraine.

3. How do you suppose a geologist can look at depositional landscape features such as a drumlin or an esker and determine it was formed from a glacier thousands of years ago?

4. Describe how ice-contact features such as kames and eskers are formed.

## Google Earth analysis

## Instructions

Fly to the following coordinates and identify the depositional landform depicted. Each of the depositional landforms listed in the key terms above will be used once. Many recognizable landforms may appear at each location, but the landforms to be identified will appear at the center of the viewing screen. Generally, depositional features are not as dramatic as erosional features, so it might be helpful to increase the elevation exaggeration to 2 or 3.

1. 43.030066, −77.573574

2. −49.442339, −73.220417

3. 43.271746, −76.878286

4. 41.710722, −70.172983

5. 45.969093, 7.816234

6. 43.024670, −77.573733

7. 49.445839, −91.143181

# 10

# ICE AGES AND INTERGLACIAL PERIODS

As pointed out earlier, glaciers and ice sheets presently cover roughly 10 percent of the Earth's surface. This, however, has not always been the case. In fact, this figure has dramatically swelled and contracted many times throughout Earth's history. The reasons for these fluctuations are complex, but all can be traced to one point of origin – Earth's climate. Earth's climate is characterized by its average temperature and average precipitation. These values, while relatively long term (tens to thousands of years), are not fixed and are subject to change. On a timescale of hundreds to millions of years, Earth's climate is prone to large swings in temperature and precipitation. Prolonged periods of global cooling and warming result in alternating cycles of freezing and thawing known as **glacial** and **interglacial periods**.

For good reason, the popular press has focused a great deal of attention on the glacier retreat of the past few decades. These changes, many scientists argue, are a product of human activity. Understanding how human activity influences global climate is inherently complex. In fact, understanding the natural variability of Earth's climate is difficult enough without adding humans into the equation. While anthropogenic causes for Earth's changing climate should not be dismissed, let's first focus our attention on the forces that control the natural variability in Earth's climate.

## Natural Causes of Climate Change

### Changes in solar output

One factor thought to govern Earth's average surface temperatures is the changes in solar output. Over billions of years solar output (irradiance) has changed. 4 billion years ago the amount of shortwave radiation reaching Earth's surface was 75 percent of its present value (Benn and Evans 2010). During the last 1,800 years alone, according to planetary scientist Robert Storm, the sun has gone through nine cycles of changes in brightness (2007). The fluctuations in solar irradiance correlate to the maximums and minimums of sunspot cycles; some scientists suggest that these changes have affected Earth's climate. For example, the "Little Ice Age" that lasted from the thirteenth century to the mid-nineteenth century corresponded to a sunspot minimum in which there was a reduction in solar radiation. During this period of sunspot minimum, global temperatures dropped significantly. The lack of sunspot activity, however, lasted for only 70 years while the "Little Ice Age" lasted for close to five hundred years (Strom 2007, 80).

## Changes in the Earth's motions

In addition to the long-term changes in output from the sun, the Earth has oscillated in its ability to receive solar radiation. Recent orbital variations over the last 2 million years have been linked to a cyclical sequence of glacial and interglacial periods. Commonly known as **Milankovitch cycles**, these orbital variations can initiate a series of cascading events that rearrange the conditions of Earth's atmosphere, oceans and of course ice.

One such variation relates to the fluctuations in Earth's orbital shape. As the Earth revolves around the Sun, it does so along an orbital plane. While many diagrams illustrate planetary orbits as circular, they are not. Often nearly circular, a planet's orbital shape can be described as an ellipse. Approximately every one hundred thousand years Earth's orbit becomes more or less elliptical. These changes in eccentricity or orbital shape respectively lengthen or shorten Earth's distance to the sun. Ultimately, these changes modulate the amount of solar radiation reaching Earth's surface. Another example of orbital variation pertains to the tilt of Earth's axis. At present, Earth's axis of rotation is tilted 23.5 degrees from the perpendicular of its plane. The inclination of this tilt fluctuates over a 41,000-year cycle, thus altering the angle and absorption of solar energy (Benn and Evans 2010).

## Plate tectonics

Aside from the external factors that manipulate Earth's climate, there are a variety of internal forces at play. The heat from Earth's core has a profound influence on Earth's climate system. Convection currents in Earth's mantle, power the tectonic processes that sculpt and reshape Earth's surface. As continents move, ocean basins develop and mountains form. Benn and Evans assert that these dramatic shifts redistribute heat across Earth's surface by restricting and

**Figure 27.** Changes in Earth's axial tilt

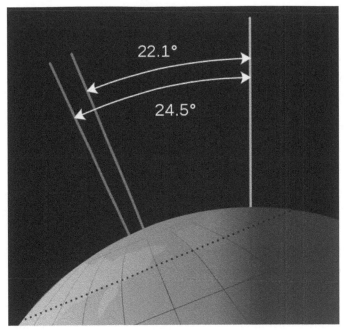

Source: NASA.

funneling oceanic and atmospheric circulation (2010, 14). Moreover, volcanoes manifest themselves at major lithospheric boundaries and hotspots where they add to the climate equation. Volcanic eruptions suspend an enormous amount of ash, dust and debris into the atmosphere; this suspended particulate matter blocks out and reflects sunlight and can result in a period of global cooling. Mini ice ages, for example, have been correlated to periods of intense volcanic activity. The eruption of the Tambora volcano in Indonesia in 1815 cooled global temperatures for several months. The following year became known as the "year without a summer" as vast regions in North America and Northern Europe experienced snow in June and July. Geologically speaking, the effects of this are short term, however, a long-term side effect may emerge.

## Changes in atmospheric composition

This long-term side effect is best explained by one of the most widely accepted theories in the atmospheric sciences – the **greenhouse effect**. Today's atmosphere is composed of nitrogen (78 percent), oxygen (21 percent) and a variety of other gases (1 percent). While that 1 percent may seem insignificant, it turns out to be very important. Many of these other gases are known as greenhouse gases. These naturally occurring gases include water vapor, carbon dioxide, methane and nitrous oxide. Like the glass in a greenhouse, these gases allow incoming solar radiation to pass through the atmosphere and reach Earth's surface. At Earth's surface, this energy is converted to longer-wavelength infrared heat where it is released back into the atmosphere. Some of this heat escapes back to space, but some of it is trapped by the greenhouse gases within the lower portions of the atmosphere and warms the Earth's surface.

Over a period of tens to hundreds of millions of years, volcanism and other geologic forces help to control the level of carbon dioxide in the atmosphere. Carbon is removed from the atmosphere in part by the weathering and burial of oceanic sediments. During periods of low volcanic activity, carbon dioxide levels are reduced as carbon is primarily sequestered in Earth's crust. A reduction in carbon dioxide translates to a reduction of global temperatures and the expansion of glaciers.

Warmer conditions return with periods of intense volcanism. The buried carbon is rereleased through volcanic activity. During the breakup of super continents such as Rodina and Pangaea for example, intense rifting and subduction contributed to volcanic outgassing, resulting in a buildup of atmospheric carbon dioxide over millions of years. Higher concentrations of carbon dioxide have been positively correlated to higher global temperatures.

## How Do Scientists Study Climate Change?

Scientists have developed a fairly detailed timeline of the sequence of Earth's ice ages and interglacial periods. This chronology starts at least 3 billion years ago and ends 10,000 years ago with the most recent ice age, the Pleistocene Epoch. Studying this past is essential in an effort to determine the future. Many scientists suggest that while the Pleistocene Epoch is over, the last ice age is not. In fact, we could be living in the midst of an interglacial interval in which the ice may return.

### Ice core analysis

To find these answers, scientists study glacial ice in an attempt to learn about past temperatures and atmospheric conditions. The ice caps in Greenland and Antarctica serve as the epicenter of this research. Scientists extract long cores of ice that read like ancient scrolls of Earth's climate history. According to Benn and Evans, the European Project for Ice Coring in Antarctica (EPICA) drilled an ice core that goes back 740,000 years, spanning the last eight full glacial cycles (2010, 52).

Air bubbles trapped in different layers of these ice cores reveal information about past atmospheric composition and temperature. Other techniques include oxygen–isotope analysis, which has been used to determine whether the snowfall has accumulated during a warm or cold period. These measurements can be made because some isotopes of oxygen are heavier than others. During warm periods, more water with heavier isotopes of oxygen evaporates into the atmosphere. Therefore the rain and snow that falls during these warm periods tend to have a heavier atomic weight than the isotopes of oxygen that fall during colder periods. Therefore, by studying the ratios of heavy to light oxygen in glacial ice, scientists can determine a sequence of glacial and interglacial periods.

## Deep-sea cores

In addition to the clues found in glacial cores, ice-age chronological data is also obtained from deep-sea cores. Sea-floor sediments contain the shells of micro-organisms that once lived at the surface of the ocean. Scientists have determined how surface water temperatures have changed over time by studying the distribution and types of shells. Likewise, the changes in oxygen–isotope ratios of micro-organisms, such as planktonic fauna, provide a record of ancient sea temperatures and can be correlated to worldwide variations in continental-ice volume. Evaporating seawater tends to leave behind more water molecules that contain heavy oxygen isotopes. During periods of intense glaciation for example, much of the water that evaporates from the ocean falls on land as snow and rain and collects in glacial ice. The net result is an increased ratio of heavy oxygen to light oxygen in seawater, which then becomes incorporated into aquatic organisms.

## Pollen evidence

This analysis is also conducted at the bottom of lakes, ponds and swamps. Pollen and evidence of fossils reside within the stratification of sediments. Such cores display detailed climate changes by revealing the types of plants equipped to deal with the climatic conditions at the time of deposition.

## Atmospheric measurements

Finally, the aforementioned data is compared to direct measurements of current atmospheric temperature, composition and trends. Temperatures are recorded from thermometers on land and at sea as well as at different altitudes with the help of weather balloons. Scientists have also employed infrared sensors on satellites to record detailed temperature readings throughout the atmosphere.

This information is then synthesized with data collected from air samples around the world that detect changes in the chemical composition of the atmosphere. Together, scientists analyze this information to understand past climate change and to predict possible future climate changes.

It is important to note, however, that like all sciences, the study of past climates has its limitations. Science cannot yield absolute certainty. The deeper one probes into the geologic past, the higher the degree of uncertainty. Ice cores and deep-sea cores, while useful repositories of paleoclimatic data, are subject to error of analysis. The lower portions of extracted ice cores, for example, are prone to ice deformation from intense pressure. Likewise, the sediments analyzed in ocean cores are often convoluted due to rates of water circulation or the biological disruption of bottom sediments. Consequently, these variations in environmental factors may lead to inaccuracies in the geologic record of climate and glacial fluctuations. Analyzing recent ice ages removes some of this uncertainty as their records are often well preserved. For these reasons, let us turn our attention to the most recent ice age – the Pleistocene Epoch.

## The Pleistocene Ice Age

During the Pleistocene Epoch, which began 1.6 million years ago, global average temperatures were approximately five degrees Celsius lower than today. While five degrees does not seem like a lot, it was cold enough to plunge much of the northern hemisphere into a "Big Chill." Up until 10,000 years ago, massive ice sheets advanced and retreated several times. Continental glaciers flowed south from Arctic and sub-Arctic regions spreading across northern North America and Eurasia. These two separate ice sheets had different points of origin but at maximum growth they fused together in a connective mass.

At its farthest advance, the North American ice sheet covered 16.2km² of land – almost 1 million km² greater than today's ice sheet in Antarctica. Divided into three parts, the North American ice sheet covered Canada, the Canadian Arctic islands, some of Alaska and central United States. The three principal parts include the **Laurentide ice sheet**, the Cordilleran ice sheet and the Ellesmere Baffin glacier complex. The latter two occupied the western and northeastern margins of the North American ice sheet respectively, while the bulk of the ice was found in the Laurentide ice sheet covering central and eastern Canada. Like the modern-day ice sheet in eastern Antarctica, the incredible thickness of the Laurentide sheet blanketed the landscape, leaving no clues at the surface as to what the underlying topography was like. It was this mass of ice that led the charge of its southern advancements.

**Figure 28.** An artist's rendition of the Earth during an ice age

Source: Ittiz.

In fact, the Laurentide ice sheet made four major advances during the Pleistocene Epoch into what is now the northern United States. These surges are named after states in which significant glacial deposits and landforms were formed during its advance. From oldest to most recent, these advancements are named Nebraskan, Kansan, Illinoian and Wisconsinan. Each glaciation during the Pleistocene Epoch was separated by interglacial intervals in which Earth's climate warmed considerably. With each new advance of a glacial surge, much of the evidence from the previous advancement was obliterated. The most recent surge, known as the Wisconsinan Stage, extended through North America as far south as southern Illinois and eastward to central Long Island.

The European theater experienced a similar glacial campaign. Like the North American progression, Europe underwent four distinct phases of Pleistocene glaciations. Each surge was separated by interglacial intervals characterized by a warmer and drier climate. The glaciers of Europe are separated into three categories – Northern Europe, the Alps and the British Isles. Similar to their North American counterparts the glacial phases that occurred in each of these regions were named after geographic locations based upon particularly vivid glacial evidence. While the European glaciers behaved similarly to those in North America, there is no definitive evidence of correlation (Sharp 1988).

## Sea-level change

At their peaks, the Pleistocene glacial advances had profound impacts on the planet. With a significant portion of Earth's water locked up in glacial ice, sea level dropped 300 feet below its current level. During glacial retreat, meltwater in conjunction with thermal expansion of the oceans flooded low-lying coastal areas. The Pleistocene Epoch helped to redraw the Earth's coastlines many times.

Glaciers initiate sea-level change in a variety of ways. In a process called **glacio-eustasy**, for example, moisture from the ocean falls over land as snow and becomes incorporated into the accumulating glacial ice. Without significant runoff from the land to recharge the oceans, global sea levels drop. When land-ice volumes decrease, meltwater returns the water to the ocean and raises global sea levels.

In addition to the global fluctuations in sea level, regional variations occur as well. Glaciers are incredibly heavy; ice sheets in particular weigh enough to depress Earth's crust and displace the underlying mantle. In a process known as **glacio-isostasy**, the crust rebounds when the load of a glacier is released as it retreats. It is helpful to visualize a sponge that has been squeezed. When the pressure is released the sponge takes back its original form. As the crust rebounds, sea level in the surrounding geographic area is usually lowered.

Evidence for this postglacial uplift can be detected within a variety of exposed landforms. Ancient formations such as raised deltas and beaches are strong indicators of former shorelines. Benn and Evans assert that estuarine flats in particular serve as exceptional locations for the study of sea-level change (2010, 235). These isolated basins accumulate a significant amount of undisturbed sediment. Within these strata of sediments, scientists collect microfossils such as diatoms. The distribution of marine sediment and fossils helps scientist reconstruct a timeline for sea-level change.

## The Great Lakes

The Wisconsinan Stage began its slow retreat to the north over twenty thousand years ago; in its wake the landscape was radically transformed. Perhaps the continental glacier's most important legacy, however, was the formation of the Great Lakes. Located in northeastern North America, the Great Lakes are a collection of five fresh-water lakes. From west to east these

**Figure 29.** The Great Lakes

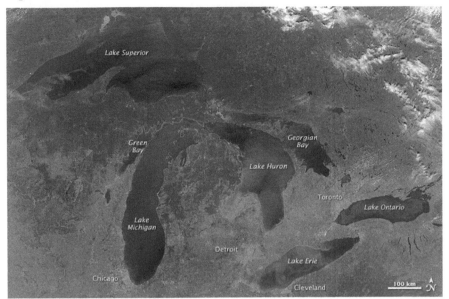

Source: NASA.

lakes are named Lake Superior, Lake Michigan, Lake Huron, Lake Erie and
Lake Ontario. While each of these lakes exists in their own drainage basin, they
are all interconnected forming a long chain, which drains into the Atlantic
Ocean to the northeast via the St Lawrence River. Hydrologically speaking,
this is a single body of water that accounts for 20 percent of the total fresh
surface water on Earth.

Although the Great Lakes are intimately connected to the Pleistocene Epoch,
their foundational geology is much older. The stage was set over one billion
years ago when the Earth was a very different place. Intense volcanic activity
laid the groundwork for the formation of the basement rocks of Lake Superior
and Lake Ontario. Lake Superior, the deepest of the Great Lakes, came into
being when Earth's crust pulled itself apart forming a midcontinent rift valley
in what is now the North American continent. The tear extends 950 miles
south from Lake Superior to Oklahoma. The rift valley was in the process
of widening to the point of creating a new ocean basin between continents
when the spreading process stopped abruptly. Evidence of this volcanism and
tectonic divergence can be found in the deepest portions of Lake Superior.
Deep-water submersibles explored the rift-valley walls at the bottom of Lake
Superior, which has a maximum depth of 406m. The basalt rocks sampled from
these depths are indicative of the volcanism that helped form the lake basin.

Lake Ontario was formed by the same set of processes. A second rift valley helped to form the Ontario basin, which extended northeast opening the St Lawrence Seaway. Both rift valleys were then overlaid by millions of years of deposition of softer sedimentary rocks.

Lake Michigan, Lake Huron and Lake Erie have very different origins. The foundational geology of these central lakes is much younger than the intense volcanism that formed Lakes Superior and Ontario. At the time of their bedrock formation much of North America was located south of the equator. Fossilized remains of ancient sea organisms and large salt deposits provide evidence that a shallow tropical-inland sea covered much of the Great Lakes region. Calcium carbonate excreted by reef building organisms built up successive layers of limestone bedrock. Much of this foundational rock, however, underwent recrystallization in the salty lagoon to form dolostone. By nature, dolostone is much more resistant than limestone, thus providing an important cap that forms the rock basin underlying Michigan, Huron and Erie. The rim of this basin forms the boundaries of the lakes. This resistant bedrock is also responsible for producing one of the most dramatic landforms in the world – Niagara Falls.

Now that the stage is set with a deep-seated geologic history, we can fast forward to the not so distant past – the Pleistocene Epoch. It is now known that the preglacial topography mimicked the current shape of the Great Lake Basin. Rivers carved their way into these landscapes following the path of least resistant bedrock. As the climate cooled and the glaciers advanced, the preexisting river valleys funneled the ice sheets leaving some areas more vulnerable to glacial action than others. The glaciers widened and deepened the valleys as they advanced and retreated many times. Eventually an ice sheet over a mile thick and 2,000 miles long overwhelmed the entire region.

Evidence of this colossal ice mass can be found littered throughout the Great Lakes region. Glacial erratics, drumlin fields and bedrock striations all point to the powerful forces of glacial erosion. Glacial deposition has left its telltale signs as well. The continental glaciers left behind a debris trail that includes till, moraines and eskers.

Approximately fourteen thousand years ago the ice sheet began to melt and retreat to the north. The meltwater pooled at the leading edge of the glacier, filling in the Great Lakes to the south; the volume of water was incredible. With the ice blocking the river outlets to the north the water flooded the Great Lakes basin producing a massive prehistoric lake known to geologists as Lake Iroquois. Engulfing the entire region, Lake Iroquois drained south into what would become the Mississippi River.

As the ice continued to retreat north, the southern shores of Lake Iroquois receded to the north as well. The ancient shorelines can still be seen south of the Great Lakes. The rolling hills south of Rochester, NY, for example, are ancient beaches of the extinct lake's shorelines. The approximate shape of the present day Great Lakes took hold 12,000 years ago. This time period coincides with the formation of Niagara Falls – an event that is indicative of the establishment of the present flow of water in the Great Lakes. The collapse of ice dams in the St Lawrence Seaway was the final obstacle removed in an effort to allow the waters of the Great Lakes to reach the Atlantic Ocean.

While the formation of the Great Lakes was complete, their morphology was not. In fact, the Great Lakes are still changing today. After the oppressive weight of the ice sheet had been lifted the land began to rebound. Similar to the loading of cargo on a sea vessel, the continental glaciers depressed Earth's crust into the underlying mantle. When the ice retreated, the land bobbed back up like a ship would when its cargo is unloaded. To the north where the glacier was the thickest, the crust was pushed 1km below normal levels; this differential resulted in dramatic differences in isostatic rebound. Where the depression was the greatest to the north, the crust rebounded as much as 150m. In many northern regions, this rebound continues today at rate of over 2.5cm per year.

Rising land translates to the lowering of water levels. As the land to the south rebounded to preglacial levels, the water reversed directions to drain out through the ice-free St Lawrence Seaway. Present-day water levels continue to drop as the land rises steadily. The effect produces yet another positive feedback loop. As lake levels continue to plummet, more weight is released and the crust continues to rebound still further. Eventually, geologists speculate that the Great Lakes will vanish altogether. This seemingly inconceivable fact speaks to the transformative effects isostatic rebound has on the landscape.

# Chapter 10 Review

### Key terms

Glacial periods

Interglacial periods

Milankovitch cycles

Glacio-eustasy

Glacio-isostasy

Laurentide ice sheet

Greenhouse effect

## Concept review and critical thinking

1. Identify and explain three natural causes of climate change.

2. In late summer 1991, Mt Pinatubo, a volcano in the Philippines, erupted and sent thousands of tons of volcanic dust into the atmosphere. Explain how the eruption of Mt Pinatubo might affect short-term and long-term climate change.

3. Identify and describe three ways in which scientists study past climates.

4. Distinguish between glacio-eustasy and glacio-isostasy. Explain how they contribute to fluctuations in sea level.

5. What role did glaciation play in the formation of the Great Lakes?

6. How does the process of isostatic rebound continue to control the morphology of the Great Lakes?

# 11

# PERIGLACIAL ENVIRONMENTS

## Definition

Adding to the already intricate assemblage contained within Earth's cryosphere are periglacial environments. The definition of **periglacial** has evolved significantly over the last 100 years. Originally, geologists restricted periglacial zones to the peripheral landscapes of the Pleistocene ice sheets. It was thought that the climate and the features adjacent to these continental glaciers were unique. However, geologist Hugh M. French argues that this narrow definition fails to recognize that the same set of processes and features occurs in vast regions that are completely disconnected and devoid of glacial activity (1996, 3). Thus, over time the definition of periglacial expanded to become more comprehensive. So comprehensive in fact that after geologists and mapmakers went back to the drawing board, they learned that the periglacial zone is much larger and more significant than they originally thought. It is now known that the periglacial domain, which includes a wide range of cold and non-glacial conditions, covers one-quarter of the Earth's land surface. All this considered, today's definition of periglacial environments are those in which the processes of intense **frost action** and activities related to **permafrost** are dominant (French 1996).

## Frost Action

Most geologists agree that frost action is the most important and universal process in periglacial environments. The term frost action actually describes the two processes of **frost heaving** and **thaw weakening**. Collectively, these processes contribute to the mechanical weathering of soil and rock material.

## Frost heaving

Frost heaving begins with the freezing of the ground. Water freezes at 0°C. Unlike most substances, liquid water becomes less dense in its solid phase. The reason for this is due to the expansion of water molecules during the freezing process, which results in an increase in total volume. Many have probably observed this phenomenon if they mistakenly leave a can of soda in the freezer too long. The result can make a mess if the expansion of freezing water causes the can to explode. Freezing water in soil, however, is a bit more complicated. In a process known as frost heave, water freezes at temperatures below 0°C. The reason for the reduced freezing point in soils is due to water's variable concentrations of dissolved salts. Likewise, water tends to freeze at lower temperatures when under pressure. As water freezes within the soil, unfrozen water exists in confined spaces; the unfrozen water creates a thin film separating the soil ice from solid particles. As pore space decreases, water's capillarity between soil particles increases. This produces a tendency of the water molecules to migrate vertically toward the surface in a process known as **cryosuction**. When the water moves up through the soil column the reduced pressure allows ice crystals to form, which displaces and heaves the overlying sediments upward.

## Thaw weakening

Thaw weakening is the second process of frost action, which has a tendency to counteract the frost-heaving capabilities of freezing ice. Seasonal warming melts some of the surface and ground ice, allowing liquid water to infiltrate the unfrozen sections of soil, further warming the material below. These thaw conditions often weaken the soil structure as water and soil settle and displace the thawed ice. This instability is enhanced when the ground refreezes. Lower temperatures create fissures due to thermal contraction in the ground. These frost cracks are then subsequently filled with ice to form an ice wedge. Such **ice wedges** may extend downward from the surface as much as ten meters. Fossilized ice wedges exist long after the ice melts as sand and gravel fill in the casts preserving their shapes.

The freeze—thaw cycles of frost action also contribute to the mechanical breakdown of rock. In a process known as cryogenic weathering, water freezes in cracks and pore spaces within rocks. As the water crystallizes, pressure builds within the rock until the expanding ice further wedges the cracks and pores open. As mentioned earlier, this process often precedes and facilitates glacial erosion. Glaciers have an uncanny ability of locating and exploiting these weaknesses in the bedrock.

This disintegration of rock varies considerably, not only with freeze–thaw patterns, but also with rock type. Some rocks, such as metamorphic gneiss, are more competent or resistant to frost action than others such as sedimentary sandstone.

# Permafrost

Despite its name, permafrost is not synonymous with permanence. Rather, permafrost is often characterized by its instability as climate change and alterations of surface conditions may result in rapid thawing. So how does permafrost differ from seasonally frozen ground? The answer resides in the duration of freezing. Seasonally frozen ground occurs where low seasonal temperatures freeze the ground only through the winter. Permafrost on the other hand can be defined as ground that remains frozen for at least two consecutive years. Often at times both conditions exist in one area; therefore the term **active layer** is used to describe the seasonally frozen ground that typically resides above the permafrost layer.

Although permafrost conditions can be found under approximately one-quarter of the Earth's land area, its distribution is highly variable both in time and space. The reasons for the variations in both the temporal and spatial distribution of permafrost are broadly related to climate. For example, while permafrost is currently forming today, a significant portion of the permafrost found in the Canadian Arctic formed thousands of years ago. Much of the icy sediment in this region has not thawed since before the Wisconsinan Stage of the Pleistocene Epoch. These icy relics of the past are not related to today's warming climatic conditions, but rather are products of a much colder climate in the recent geologic past. Globally, however, permafrost is declining. If the current warming trend continues the permafrost's southern boundary will continue to retreat poleward. The ecological consequences of significant permafrost thawing due to global warming are discussed in the next chapter.

Spatially, permafrost exists in areas of high latitude and high altitude where the annual air temperatures are low. The climate parameters required for the formation of perennially frozen ground, however, are extremely variable. Cold-air temperatures are of course a prerequisite for the formation of permafrost. Geologist Neil Davis goes a step further by outlining the critical climate conditions required for permafrost formation (2001). As a general rule, mean air temperatures near 0°C will produce scattered patches of permafrost. These isolated and discontinuous layers of permafrost are typically found below an active layer that is 1–2m thick. As latitude or altitude increases and average temperatures are several degrees below 0°C, the active layer thins and permafrost thickens. Under these conditions, the permafrost layer is still discontinuous but its thickness often measures between 30 and 50m. Travelling farther north to various latitudes in Siberia for example, the permafrost layer extends to even greater depths. In these regions, the active layer thins to only 1–2cm, while continuous permafrost extends down to a depth of 400m. The deepest recorded depth of Siberian permafrost reached an incredible 1,450m (Davis 2001).

Air temperatures alone, however, are of limited use when predicting local variations in permafrost. Instead, scientists often rely on mean average ground temperature. Permafrost requires average ground temperatures below 0°C. Even with these conditions, permafrost is not always guaranteed. Moreover, if permafrost is found, its distribution is not always uniform. For this reason permafrost is often classified as either continuous or discontinuous. Continuous permafrost indicates that perennially frozen ground is found at all locations with the exception of localized thawed zones such as areas adjacent to bodies of water. Discontinuous permafrost on the other hand occurs in pockets where frozen layers in the ground are separated by layers of unfrozen ground. The reasons for these variations in distribution are complex, but many of them can be related to terrain conditions. French argues that topographic relief, aspect, vegetation, snow cover and presence of water have strong controls on ground conditions and therefore extent of discontinuous permafrost (1996, 59).

Both topographic relief and aspect influence the amount of solar radiation received by the ground as well as the amount of snow accumulation. Variations in solar radiation and snow cover are particularly well documented in alpine regions in the Northern Hemisphere. Highland environments with different slope orientations for example, receive different amounts of solar radiation. For instance, a south-facing slope will experience more solar radiation and therefore less snow cover than a north-facing slope. Mountainous regions in the Yukon exhibit this relationship as permafrost is common on the north-facing slopes, but is virtually nonexistent on the not too distant south side (French 1996).

In addition to topography, vegetation strongly influences the extent of permafrost. In general, vegetation shields the underlying layers of permafrost from the warming potential of solar radiation. The forested taiga, in particular has profound effects on the local permafrost conditions. Depending on the tree species within these boreal forests, more or less shade is provided. The spruce forests of North America, for example, offer more shade and tend to have a thinner active layer than the less dense canopies of the pine and tamarack forests common in Siberia (French 1996).

Trees also help to capture snowfall. The less snowfall that accumulates on the ground, the deeper those colder temperatures can penetrate. Deep snow cover acts as a blanket of insulation that can prevent frost penetration. Yet, these relationships between tree and snow cover are not set in stone. Upland areas without trees tend to experience harsh winds, which can remove snow cover and promote the formation of permafrost. On the contrary, the longer that snow cover persists into the spring and summer months, the less thawing of the active layer that will occur.

Another critical factor controlling the distribution of localized permafrost is water. Of all the substances found on planet Earth, liquid water has the highest specific heat. This means that water requires the most energy per gram to heat up, in other words, water is a poor absorber and radiator of heat energy. As a result, **taliks**, or unfrozen layers of ground often exist near and underneath bodies of water that do not completely freeze during the winter months.

Beside the factors outlined above, the composition of the soil and rock contribute to the distribution of permafrost. The soil matrix in which permafrost exists is rarely a homogenous unit. Multiple minerals may heat and cool at different rates and therefore complicate the rates of freezing. In addition to the variations in thermal conductivity, there are the diversities of albedo for different minerals. The reflectivity of different rock-forming minerals can range from 10 to 40 percent and therefore influence the extent of permafrost and depth of the active layer (French 1996, 77).

# Periglacial Landforms

Like glacial erosion, the processes that dominate the periglacial domain produce unique features in the landscape. While these landforms are rarely as spectacular and dramatic as those produced by glacial erosion, periglacial environments are just as distinct and sometimes even more peculiar.

## Pingos

French contends that many surface features in periglacial environments manifest themselves in response to both the aggradation and degradation of permafrost (1996, 51). As discussed earlier, the inherent instability of permafrost modifies the surface as it builds and subsides. **Pingos**, for example, are ice-cored conical-shaped hills commonly found in the midst of continuous permafrost. The injection and subsequent freezing of pore water from a nearby talik build up these dome-shaped mounds of soil and ice. Pingos range in height from a few

**Figure 30.**  Pingos near Tuktoyaktuk, Northwest Territories, Canada

Source: Emma Pike.

meters to 60m and up to 300m in diameter. Some pingos have depressions or cracks at their top caused by the melting of the ice core or a buildup of water pressure. Several Russian scientists have actually witnessed the violent rupturing of a pingo. Resembling a small volcano, the explosion expelled large blocks of ice into the air and spouted pressurized water for 30 minutes (Davis 2001).

## Thermokarst

Juxtaposing the buildup of permafrost is of course the melting of it. Such degradation of permafrost might be attributed to a variety of factors including changes in climate or an alteration in vegetation conditions. As the bodies of ice in permafrost melt, the voids left behind often destabilize the soil structure. These failures in soil structure cause the ground surface to collapse and subside leaving behind a topography characterized by pits and depressions. This type of terrain is referred to as **thermokarst**. Thermokarst processes, according to French "are some of the most important processes fashioning the periglacial landscape" (1996, 109). Indeed, the thaw lakes and depressions commonly found in the lowland areas of Alaska's North Slope are quintessentially periglacial in nature. These shallow, elliptical depressions also known as tundra ponds are part

**Figure 31.** Thermokarst lakes on Alaska's North Slope region

Source: NASA.

of thermokarst terrain. Both satellite and aerial imagery depict these vast regions in Alaska as landscapes that resemble swiss cheese, as the water-filled depressions look like holes in the land. Thermokarst phenomena such as these thaw lakes are products of melting, subsidence and accumulation of water in the depression.

## Patterned ground

The notable landforms of periglacial environments are not limited to the processes related to permafrost. Cryogenic processes – those that involve repeated freezing and thawing of the active layer, also produce striking features in periglacial landscapes. Among the most fascinating landforms produced from these processes is **patterned ground**. At first glance these systematic arrangements of microrelief look as though humans created them. Some arrays

**Figure 32.** Patterned ground structures on the Svalbard Archipelago

Source: Hannes Grobe.

possess a distinctive symmetry that is comparable to an artistic pattern or tiles on a kitchen floor. Nevertheless, patterned ground is a naturally repetitive feature commonly found in the periglacial domain. Patterned-ground phenomena express themselves as topographic features or as differences in composition and sorting of material at the ground surface. Thus, patterned ground is classified based on its geometric form and the presence or absence of sorting.

The geometric forms of patterned ground can be subdivided into two broad categories. Category one is described as closed patterned and may include interconnected geometric shapes such as polygons and circles. Closed patterns such as these are typical of flat or nearly flat topography. These patterns occur in regions exposed to moisture and seasonal freezing and thawing cycles. Polygon shapes are most likely created from similar processes that instigate frost heaving. Large rocks and boulder are thrust vertically through the soil to the surface. Geologist Jon Erickson suggests that large rocks move through the soil "by a pull from above and a push from below" (1996, 219). In other words, in conditions where the top of a large rock freezes first, the expanding frozen soil pulls it upward. During periods of thaw, sediment gathers below the rock causing it to settle at a higher level. Furthermore, the expanding freezing subsoil helps to heave the rock up as well. Collectively, these forces over consecutive frost–thaw cycles can bring large masses of rock to the surface (Erickson 1996).

Sorted stone circles and sorted stone polygons are perhaps the most mesmerizing forms of patterned ground. These arrays consist of a central area of fine sediment grading out to courser rocky material. According to Davis, differential frost heave plays the major role in forming these geometric patterns. These arrangements suggest that the soil has been churned up through a variety of cryogenic processes that produces what some scientists refer to as convection cells in the soil. For example, a process known as **cryoturbation**, creates circular forms of patterned ground. This sequence of events refers to the vertical and lateral soil movements that occur in response to ground displacement associated with frost action. Some circular patterned forms are slightly dome shaped which help signify the underlying process of vertical heaving and churning of soil (French 1996).

Opened patterns are the second broad group of geometric shapes categorizing patterned ground. Steps and stripes are considered open patterned and are commonly found in sloped periglacial environments. Geologist A. L. Washburn claims that as the surface gradient increases, circular- or polygon-shaped patterned-ground features become elongated due to the gravitationally induced downslope movement of debris (1973, 101). On steeper-sloping hillsides, a series of small benches known as steps may emerge. Terrace-like steps run perpendicular to the slope and consist of a thread or bench and a riser of roughly the same size. Over time, continued mass wasting will reduce the slope angle and produce ground patterns that present themselves as parallel or semiparallel bands of material in a striped formation running with the slope. Mass wasting sorts the debris within the stripes according to the material's rate of downslope movement.

## Chapter 11 Review

### Key terms

| | |
|---|---|
| Periglacial | Active layer |
| Frost action | Talik |
| Permafrost | Thermokarst |
| Frost heaving | Pingos |
| Thaw weakening | Patterned ground |
| Ice wedge | Cryoturbation |
| Cryosuction | |

*Concept review and critical thinking*

1. Define periglacial and identify the geographic extent of the periglacial domain.

2. Distinguish between frost heaving and thaw weakening and explain how they contribute to the mechanical breakdown of soil and rock.

3. What is permafrost? What factors control whether permafrost is continuous or discontinuous?

4. Compare and contrast pingos with thermokarst.

5. Explain how patterned ground forms. Distinguish between closed and open patterns and explain how each arrangement emerges.

# GLACIERS AND GLOBAL WARMING

Climate change is not new. The myriad of glacial and interglacial periods over the last 900,000 years speaks to the long-term variability of the Earth's climate system. However, there is much scientific evidence to suggest that the Earth is currently warming at an accelerating rate. This warming is best expressed by the melting of glaciers. In 2002, satellite imagery captured the collapse and subsequent disintegration of the Antarctic Peninsula's Larsen-B ice shelf (Figure 1). Many scientists contend that this event is a symptom of a larger ailment that is inflicting the Earth. This ailment is of course global warming.

## Anatomy of the Atmosphere

Before delving into the implications of human-induced climate change and its long-term effects on the Earth's cryosphere, let us first develop an understanding of the basics of atmospheric science. Earth's **atmosphere** is extremely delicate. Bound to the Earth by gravity, the atmosphere is a spherical blanket of gases that envelopes the Earth. Relatively speaking, the atmosphere is very thin. A simple analogy helps to illustrate this point; if the Earth were the size of an apple, the skin of the apple would represent the thickness of the atmosphere.

In keeping with food analogies, like lasagna, the atmosphere is layered. Each layer is defined by marked changes in temperature. Extending from Earth's surface to outer space these layers include the **troposphere**, stratosphere, mesosphere and thermosphere. Since gravity pulls the atmosphere toward Earth's surface, the majority of the gas molecules are found in the lowest layer or troposphere. As a result, the troposphere is the most active layer of the atmosphere. The troposphere is also the only layer that contains water vapor, which means it is the only layer in which weather occurs.

# The Greenhouse Effect

The troposphere is composed primarily of two gases – nitrogen (78 percent) and oxygen (21 percent). This leaves 1 percent to other gases. It turns out, however, that this one percent is very important. Some of these other gases include carbon dioxide, water vapor, methane and nitrous oxide. Collectively, these compounds are known as Earth's greenhouse gases. Without these gases, life on Earth would not be possible. The reason these gases are so important is because they help warm Earth's lower atmosphere through a natural process known as the **greenhouse effect**. The greenhouse effect owes its name to the warming effect produced inside a greenhouse. The panes of glass covering a greenhouse allow incoming sunlight to pass through. The visible light is then absorbed and reradiated as longer-wavelength heat energy, which is then trapped inside the greenhouse by the glass. Earth's troposphere operates in this same way. Greenhouse gases trap outgoing infrared energy warming the lower portions of the atmosphere. Without this warming effect, the Earth would in fact be a frozen wasteland. Thus, the greenhouse effect has become one of the most widely accepted theories in the atmospheric sciences.

# Global Warming

Many scientists contend that human activities are enhancing the natural greenhouse effect and ultimately raising global temperatures. This human-induced climate change has been termed **global warming**. Since the industrial revolutions of the eighteenth and nineteenth centuries, humans have fundamentally transformed the composition of Earth's atmosphere. Human actions such as the combustion of fossil fuels, deforestation and agriculture have increased the concentrations of greenhouse gases including carbon dioxide, methane and nitrous oxide.

### Carbon dioxide

The most important greenhouse gas that is directly influenced by human activities is carbon dioxide. Part of the global carbon cycle, carbon dioxide provides the primary means by which carbon is transferred through the environment between several carbon reservoirs. The major reservoirs of carbon include the biosphere, hydrosphere, atmosphere and lithosphere. Terrestrial producers remove carbon dioxide from the atmosphere, likewise, aquatic producers remove dissolved carbon dioxide from water. Through the process of photosynthesis, these producers convert carbon dioxide into complex carbohydrates such as glucose. In another process known as aerobic respiration, the glucose is then broken down by producers, consumers and decomposers, which in turn releases carbon dioxide back into the atmosphere.

The timescale of carbon circulation is highly variable. The turnover time for carbon exchange in the top layers of the ocean and terrestrial forests, for example, can range anywhere from less than a year to several decades. Some carbon atoms, however, take much longer to recycle. They can become trapped for millions of years in the carbonate rocks of deep ocean basins or beneath repeated accumulations of organic material in tropical swamps. These sequestered carbon atoms would never see the light of day unless intense volcanic upheaval violently forced them back to the Earth's surface. This would all change of course. While under intense heat and pressure over millions of years, many carbon atoms morphed into precious commodities – fossil fuels. The carbon found within coal, oil and natural gas would soon be brought back to Earth's surface through human excavation and then burned.

In only a few hundred years human activity has fundamentally altered the carbon cycle. Ice-core analysis has revealed that prior to the industrial revolutions of the eighteenth and nineteenth centuries, the exchange of carbon dioxide between its major reservoirs was relatively constant. According to scientist John Houghton, the atmospheric concentration of carbon dioxide was kept within about 20 parts per million (ppm) of a mean value of about 280 ppm (2009, 31). Detailed measurements now reveal that atmospheric carbon dioxide concentrations have risen from 285 ppm in 1850 to 400 ppm in 2013. This rise in concentration in carbon dioxide is directly attributable to human activity. The largest contributor by far is the extraction and burning of fossil fuels for human energy needs. Combusting carbon-based coal, oil and natural gas releases vast quantities of carbon dioxide into the atmosphere.

Another human activity contributing to the increase in atmospheric concentration of carbon dioxide is deforestation. The cutting and burning of forests not only releases carbon dioxide but it also eliminates important reservoirs of carbon dioxide. Forests are what scientists call carbon sinks. Through photosynthesis, trees sequester carbon dioxide – limiting atmospheric concentration. Thus, fewer trees mean more carbon dioxide. These disruptions to the carbon cycle have profound implications for Earth's climate.

## Methane

Another greenhouse gas derived from human activities is methane. While less abundant than carbon dioxide, methane is decidedly more potent in terms of its heat-trapping capabilities. The human activities responsible for releasing

**Figure 33.** Deforestation in the Amazon interior from 1992 (top image) to 2006 (bottom image)

Source: NASA.

methane include raising cattle, flooding land to create reservoirs, creating landfills and extracting and transporting fossil fuels. Ice cores have revealed that because of these activities, methane concentrations have more than doubled since the year 1800 (Houghton 2009). Methane also exists naturally in the ice-encased organic material locked up in the Arctic's permafrost. In yet another positive feedback loop, methane emissions are expected to rise in the future as global warming continues to melt the permafrost found in the Arctic.

### Nitrous oxide

Nitrous oxide accounts for approximately 9 percent of the greenhouse gas emissions from human activity. Like methane, nitrous oxide has very high warming potential – about two hundred and thirty times that of carbon dioxide. The major anthropogenic (manmade) sources of nitrous oxide are inorganic fertilizers. Since the advent of modern agriculture, ice core analysis has shown a 16 percent increase in the nitrous oxide concentrations (Houghton 2009).

## Predicting Climate Change in the Twenty-first Century

According to the 2001 report of the Intergovernmental Panel on Climate Change (IPCC), by the year 2100, global temperatures will rise between 1.4°C and 5.8°C. These projections are based in part on the likely global emissions of greenhouse gases. However, understanding how these changes in climate will be distributed across the globe is much more complicated; there is significant variation in the way the Earth warms. What happens at the poles, for example is decidedly different from what happens at the equator. While climate models are becoming more sophisticated and able to account for an amazing array of variables, uncertainty still exists.

The vast majority of these uncertainties pertain to how the Earth will respond to warming temperatures. Many climatologists anticipate a series of feedback loops, both positive and negative. Scientist Sir John Houghton suggests that there are four major feedback loops that are expected to continue in response to a warming world. These feedbacks include water-vapor feedback, cloud-radiation feedback, ocean–circulation feedback and ice-albedo feedback (2009, 90).

### Water-vapor feedback

Water-vapor feedbacks are positive feedback loops that occur as warmer temperatures increase the rate of evaporation from the oceans and wetland surfaces. Water vapor is a potent greenhouse gas, and therefore as the atmosphere becomes moister, global temperatures continue to rise. Many climate models estimate a doubling of atmospheric temperatures when compared vis-à-vis with models of fixed water-vapor concentrations.

## Cloud-radiation feedback

There is also uncertainty revolving around the effects of cloud cover on atmospheric warming. In fact, cloud-radiation feedbacks may produce one of two polarized outcomes in which the atmosphere either cools or warms. Scenario one results in a cooler world. As previously mentioned, an increase in atmospheric temperatures increases the amount of water vapor in the atmosphere, which in turn creates more clouds. In general, an increase in continuous low-altitude clouds reduces the amount of solar energy reaching Earth's surface by reflecting it back out to space. Scenario number two results in a warmer world. An increase in thin high-altitude clouds allows the solar energy to reach Earth's surface, but prevents the reradiated heat from escaping. Scientists are still hard at work trying to determine which one of these effects will be most dominant.

## Ocean-circulation feedback

Oceans help to moderate Earth's climate. They do this by serving as the major source for atmospheric moisture, absorbing heat from the atmosphere, and redistributing heat through their circulation patterns. These factors moderate Earth's climate by reducing the extremes of atmospheric temperatures. For example, coastal climates tend to have cooler summers and warmer winters. Average monthly temperatures of inland areas on the other hand experience dramatic swings. Similarly, latitudinal variations in atmospheric temperature are modified in part by the transport and redistribution of heat through ocean currents. Therefore, the rate of atmospheric change is and will be largely controlled by the oceans. Another ocean–circulation feedback that is expected to influence the rate of atmospheric change is related to the ocean's ability to absorb and recycle carbon dioxide. Currently, the oceans absorb 25–30 percent of human emissions of carbon dioxide from the lower atmosphere. However, as ocean temperatures increase, their abilities to absorb carbon dioxide are diminished. Further compounding this problem is the fact that as ocean temperatures increase, more carbon dioxide is released from the oceans to the atmosphere. Thus, another destructive positive feedback loop emerges where higher temperatures increase atmospheric carbon dioxide levels, which in turn further increases atmospheric temperatures.

### Ice-albedo feedback

As mentioned earlier, the variations in warming across the Earth will not be uniform. Most climate models predict that the greatest change in atmospheric conditions will occur in the polar regions. Ice-albedo feedbacks are a major contributor to this regional variation. Regions covered with light-colored snow and ice are typically cool in part because of their high albedo. Albedo is a measurement of reflectivity. Snow and ice reflect more sunlight than darker-colored surfaces such as land and water. As global temperatures rise and the snow and ice start to melt in the polar regions, a greater portion of darker land surfaces and sea area will be exposed. The reduction in albedo will lead to more atmospheric warming as more solar radiation is absorbed at Earth's surface. In yet another feedback, snow and ice continue to melt and temperatures continue to warm. Climate models predict that ice-albedo feedbacks on their own could warm global temperatures by 20 percent in a world with double the carbon dioxide (Houghton 2009).

# Projected Outcomes of a Warmer World

The efficacy of computer models in projecting future climate change is in large part determined by their ability to account for the aforementioned feedbacks. Further adding to this complexity, computer-model simulations need to be validated against observations of past and current climate change, climatic cycles such as El Niño events, and atmospheric changes due to a variety of anomalies such as volcanic eruptions (Houghton 2009). Needless to say predicting future climate change is very complicated.

Complications, notwithstanding, the Earth is warming. As mentioned earlier, projections indicate that in a world with increased greenhouse gas emissions, the global average temperature will rise between 1.4°C and 5.8°C. Scientists are not only concerned about how much temperatures change, but perhaps more importantly about how rapidly these changes are taking place. Like any natural system, climate systems have tipping points. A tipping point is a threshold level at which change to a system becomes irreversible. Time delays in a system often mask the changes taking place and reduce the effectiveness of a negative feedback loop's counteractive measures.

Many scientists contend that several climate-changing tipping points are embedded in the cryosphere. For example, it has been suggested that the collapse and melting of the Greenland ice sheet would lead to large-scale, irreparable and potentially catastrophic climate changes. Other such tipping points include the

**Figure 34.** Alpine glacial recession: The retreat of the Gangotri Glacier in India

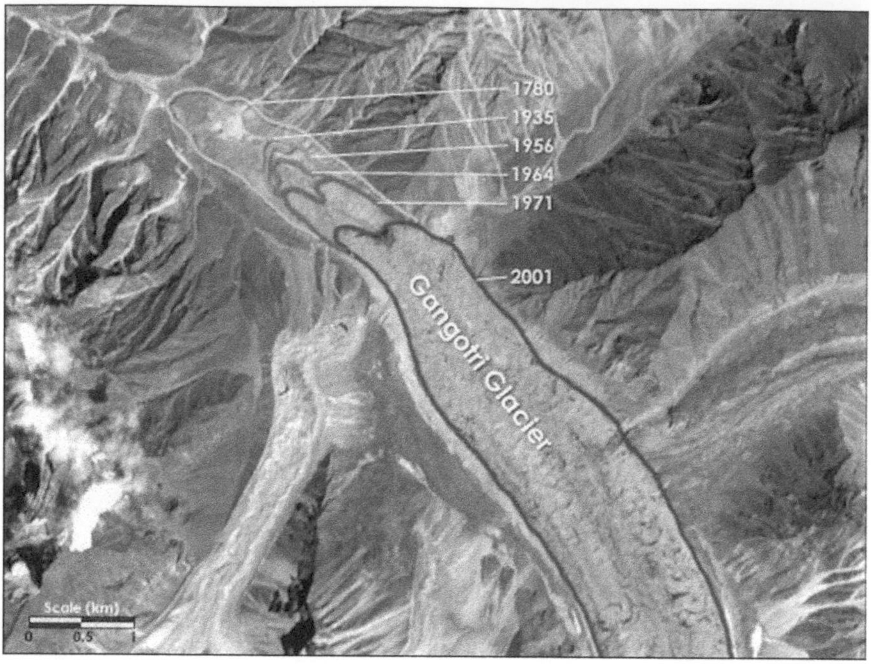

Source: NASA.

collapse and melting of the western Antarctic ice sheet, melting of the Arctic-summer sea ice and widespread thawing of the Arctic permafrost. Scientists urge us to avoid crossing such thresholds, yet we still don't know how close we are to reaching these tipping points. What we do know is that much of Earth's ice is melting at an unprecedented rate.

## Global warming and alpine glaciers

Mountain glaciers exist at all latitudes and are therefore probably the most extensively studied glaciers in the world. The effects of global warming on these vital systems are well documented and alarmingly consistent. The scientific consensus indicates that most alpine glaciers are receding. To be fair, the recession has not been uniform, and these trends have been interrupted by readvances. Several mountain glaciers in Norway and Switzerland, for example, have defied the trends by advancing in response to increased winter precipitation. Nevertheless, the overall pattern is one of intense melting.

During the last 25 years many of the world's alpine glaciers have not only been shrinking, but have been receding at an accelerating rate. According to the

**Figure 35.** An enormous piece of ice, roughly 250km² in size, breaking off from the Petermann Glacier along the northwestern coast of Greenland in 2010

Source: NASA.

World Glacier Monitoring Service, there has been a cumulative net loss of 9m thickness for 30 glaciers in nine regions of the world since 1985 (Strom 2007, 152). An even more inclusive study revealed that since 1988, the rate of ice loss for all the world's glaciers has more than doubled. At this current rate, it has been estimated that all alpine glaciers will be gone in about seventy-five years (Strom 2007).

The consequences of this accelerated melting will be far reaching. Alpine glaciers play a pivotal role for many mountain communities. People living in South America's Andes range for example, rely on an ever-dwindling supply of glacial meltwater. As global temperatures rise, snowmelts occur earlier in the spring. This realigns peak river runoff with winter and early spring and away from summer and autumn when demand is the highest. The results could affect millions of people living in countries such as Ecuador, Bolivia and Peru who rely on the meltwater for irrigation and hydropower.

## Global warming and continental glaciers

As pointed out in Chapter 3, there are only two ice sheets in the world. One ice sheet covers most of Greenland near the Arctic Circle, while the other ice sheet covers the Antarctic Continent at the South Pole. Unfortunately for both of them, climate models predict that the most severe climate change will occur in the world's polar regions.

## Greenland

According to the 2007 IPCC report, over the last 50 years Arctic temperatures have risen twice as fast as the average temperatures in the rest of the world. Moreover, current projections indicate that the Arctic will experience more warming than the rest of the world over the next 100 years. This accelerated warming in the Arctic is in part a function of ice-albedo feedbacks. Arctic-summer sea ice is decreasing at a rate of 8 percent per decade (Strom 2007). As the ice melts, it is exposing more dark surfaces causing the ice to melt further. Studies indicate that the summer sea-ice coverage will be gone by 2040 or earlier.

This Arctic-heating trend is having profound consequences on Greenland's ice sheet. The effects, however, are paradoxical. The interior of Greenland's ice sheet is actually building with snow accumulation while the margins of the ice sheet are melting at accelerated rates. Interestingly, both results are thought to be products of global warming. The interior of Greenland's ice sheet is thickening by 6.4cm per year above an elevation of 1.5km per year (Johannessen et al. 2005). As atmospheric temperatures increase, evaporation from the warming ocean increases as well. The warmer air has higher water-vapor holding capacities and therefore precipitation in the interior of Greenland increases. Meanwhile, along the periphery of Greenland's ice sheet, melting is the dominant activity. Satellite imagery reveals an 18 percent expansion of the melt region between 1992 and 2005 (Strom 2007). Even more alarming is that over this time period the rate of melting accelerated. Between 1992 and 1997 the average ice loss was 60km$^3$ per year, but between 1997 and 2003 the ice loss increased to 80km$^3$ per year (Krabill et al. 2004).

The reasons for this rapid acceleration have to do with the mechanics of a melting ice sheet. The leading edges of the largest glaciers in Greenland are massive. Many of these glacial tongues extend out into the sea, partly submerged in the water. These ice shelves serve as critical barriers for the ice found in the region's interior and are the first line of defense against global warming. As a result, ice shelves are extremely vulnerable as they are exposed from above and below. From above, warm air melts the ice shelf creating massive pools of meltwater on the surface. Many of these meltwater pools bore down into the glacier and pour down into moulins, which are deep holes that can eventually reach the base of the bedrock. From below, warmer ocean water flows and burrows into cavities beneath the ice shelf. Together with the meltwater from above, basal melting helps to lift the ice sheet off the bedrock. The destabilized ice shelf weakens and cracks. Eventually, the ice shelf starts to collapse as massive chucks of ice calve into the sea. Like letting the cork out of

**Figure 36.** Mass balance atmospheric circulation

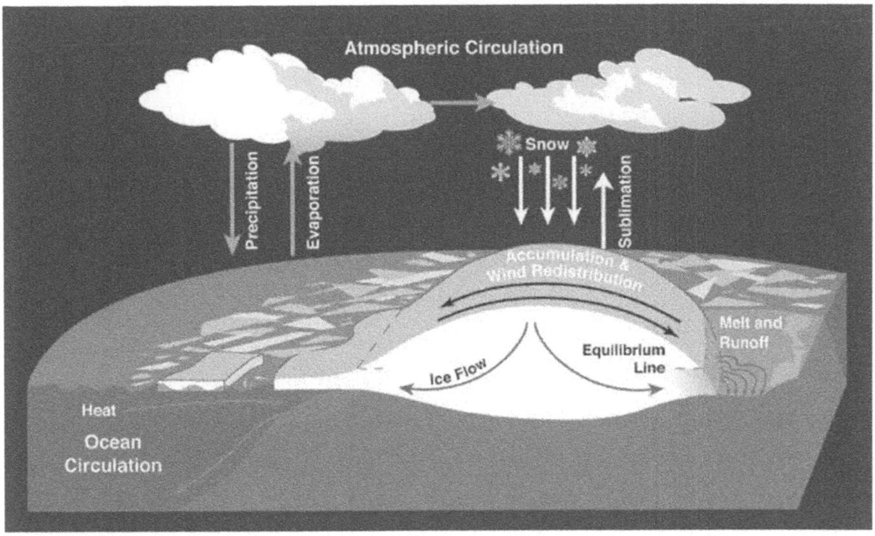

Source: NASA.

a champagne bottle, once the ice shelf is gone the unrestrained glaciers behind it slip into the sea.

## Antarctica

The Antarctic ice sheet is the largest mass of ice on Earth. Located at the South Pole, the ice sheets of Antarctica comprise 90 percent of all ice on Earth. Like in Greenland, the effects of global warming in Antarctica are contradictory. As temperatures warm, there is more evaporation from the oceans and more precipitation in the interior of the continent – particularly in the eastern part of Antarctica. Over the last 25 years, snow accumulation in these regions has actually increased. Accumulation, notwithstanding, detailed satellite measurements show that there has been a significant overall decrease in the ice sheet's mass.

The majority of the melting is happening along the western portion of Antarctica's ice sheet. Most of the erosion by melting is occurring along the boundaries of ice sheets located on the western side of the Antarctic Peninsula. Like the ice shelves found along the periphery of Greenland's ice sheet, Antarctica's ice shelves act as a critical barrier preventing the advance of the continental glaciers toward the sea. In March 2002, a significant portion of this barrier collapsed and disintegrated into the sea. Nearly two-thirds of Larsen-B, an ice shelf the size of Delaware succumbed to years of melting.

This catastrophic event finally put the uncorked-champagne-bottle hypothesis to the test. In summary this hypothesis states that if a major ice shelf is removed, then the unrestrained continental glaciers once behind the ice shelf rush into the sea. The results of the Larsen-B collapse quickly validated the hypothesis as many of the glaciers behind the ice shelf rapidly moved into the sea. In what many scientists refer to as unequivocal evidence of global warming, the major glaciers behind the Larsen-B ice shelf not only slipped into the sea, but also began to disintegrate inland. The notion that the centers of ice sheets are invincible had to be discarded. It turns out that the loss of the peripheral ice shelves makes the entire ice sheet vulnerable to extinction.

## Global warming and rising global sea levels

The United States Geological Survey concluded that global sea levels will most likely rise between 0.8–2m by the end of this century. If these estimates are correct, over 150 million people could be turned into refugees as rising sea levels spill over populated coastlines. Many of the developing countries of the world will experience the greatest brunt of this potential cataclysm. Bangladesh, a country of 160 million people, for example, is expected to experience regular flooding and inundation of over half its land area. Countries such as these are especially at risk because they lack the resources to reduce the impacts of coastal flooding or to relocate large populations of stranded people. To make matters worse, many of these countries are located in areas of the world that will not only experience a rise in sea levels but a constant bombardment of high-intensity hurricanes and typhoons. In fact, the very existence of several island nations in the Pacific is at stake as rising sea levels and severe storms threaten to wash them away.

These changes are of course not limited to the developing world. Recent research suggests that the northeast coast of the United States is likely to experience the largest sea-level rise in the world. Cities such as Boston and New York would be among the hardest hit by these dramatic transformations. Aside from the obvious threats of storm damage and the inundation of low-lying coastal areas, there are a series of other effects of rising sea level that threaten ecosystems and public health. The ecological consequences of rising sea level include increased erosion of beaches and the destruction of critical habitat such as coastal marshes. Human health comes into question when saltwater pushes inland and contaminates coastal aquifers. The effects of these impacts can destabilize the very economies that coastal communities are built on.

Scientists assert that the projected rise in sea levels is directly attributable to global warming. Ice-core analysis reveals that there is a strong correlation between sea-level rise and the atmospheric content of carbon dioxide. Throughout the last ice age, carbon dioxide levels helped to regulate atmospheric temperatures. During the maximum extent of glacial ice 21,000 years ago for example, carbon dioxide levels were at their lowest – approximately 185 ppm. During this time, glacial ice covered a significant portion of North America and Europe and sea levels were 130m lower than today. Compare these values to the prevailing conditions 35 million years ago when the average global temperature was 12°C warmer than today. Carbon dioxide levels were a staggering 1,250 ppm which translates to a sea level 73m higher than today (Strom 2007). Many scientists contend that the recent warming trends related to human activity are triggering these changes today.

Global warming contributes to a rise in sea level through two mechanisms. Mechanism one is the thermal expansion of the oceans. As the oceans warm, the seawater expands and sea levels rise. Because of differences in heating of the different parts of the world's oceans, sea-level rise due to thermal expansion is not uniform. Nevertheless, scientists estimate that thermal expansion of the oceans, accounts for at least half of the observed rise in sea level worldwide.

The second reason why sea levels rise as Earth's temperatures warm is due to the large inputs of fresh water into the oceans from melting ice. To be sure, not all ice melting contributes to a change in sea level. For example, there is a difference between sea-based ice and land-based ice. The disintegration of sea-based ice has little effect on sea level as the meltwater simply displaces the same volume of water occupied by the ice mass. Melting land-based ice on the other hand discharges large volumes of fresh water into the oceans and therefore can significantly raise global sea levels. Therefore, much attention has been directed toward the land-based glaciers of Greenland and western Antarctica as well as the major alpine glaciers of the world.

In particular, the accelerated discharge of the outlet glaciers of Greenland and the rapid disintegration of western Antarctica's ice shelves is cause for concern. Nevertheless, there is much uncertainty when it comes to predicting the effects of global warming on future sea levels. According to some models, it will take a rise of more than 5°C to instigate a total meltdown of the Antarctic ice sheet (Hambrey and Alean 2004). Taking it to this extreme, if the entire ice sheet was to melt under these conditions, it would add approximately 30 million km$^3$ of water to the world's oceans, raising sea levels by about 60–70m (Hambrey and Alean 2004). Admittedly, this scenario is not likely to happen in the near future, and while these values seem astronomical and are perhaps at the extreme end of a longer-term spectrum, they do illustrate an important point. That point being that even a smaller percentage change in the Antarctic ice sheet could have profound effects on global sea level.

Although these major changes in sea level appear long term, there is measurable change in motion today. Melting glaciers along the western coast of Antarctica for instance, have raised global sea levels between 0.13–0.24mm per year since the mid 1990s (Shepherd et al. 2004). Likewise, the Greenland ice sheet, a continental glacier even more vulnerable to climate change than the Antarctic ice sheet, is raising global sea levels as well. Scientists have estimated that breakup and subsequent melting of some of the peripheral glaciers in Greenland may have contributed to a sea-level rise of up to 0.09mm per year since the 1990s (Joughin et al. 2004). Further adding to the incredible volumes of discharge melting into the sea from ice sheets are the meltwaters flowing from alpine glaciers. The accelerating rate of recession of mountain glaciers is estimated to account for about half the global rate of ice loss and is thus a major contributor to rising sea levels. Melting alpine glaciers in Alaska and northern Canada for example, have raised global sea levels by 0.32mm per year over the last ten years (Hambrey and Alean 2004).

## Global warming and periglacial environments

The enhanced atmospheric concentrations of greenhouse gases will have profound consequences on periglacial environments. As global temperatures rise due to global warming, the extent of seasonal snow cover, ground ice and permafrost is expected to decrease. The loss of permafrost is of particular importance because as it thaws, increased bacterial action in the soil releases methane, a greenhouse gas locked up in frozen, decomposing organic matter. As mentioned earlier methane is a potent greenhouse gas that is about twenty-five times higher in its thermal capacities than carbon dioxide. These emissions of methane may therefore facilitate destructive positive feedback mechanisms that could dramatically accelerate atmospheric warming. For example, western Siberia has warmed faster than almost anywhere on Earth. Temperatures have increased by 3°C over the last 40 years (Strom 2004). The result is a significant thaw of Siberia's periglacial domain. Scientists estimate that the permafrost in western Siberia alone contains approximately 70 billion tons of methane. Much of this methane is expected to release within the next few decades, further compounding the destructive positive feedback loops where accelerated atmospheric warming thaws more permafrost, which then releases more methane. Major thawing is also taking place in Alaska, Canada and Scandinavia. As the degradation of permafrost occurs across all of these periglacial domains, scientists anticipate a 0.4°C increase in global temperatures by 2020 and perhaps a 0.6–0.7 increase by 2050 (Street and Melnikov 1990).

Aside from the positive feedback mechanisms, melting permafrost has several other consequences. Thawing permafrost threatens entire forested ecosystems as the ground in which the trees are rooted in become unstable. The trees in these drunken forests are displaced from their vertical alignment and are often tilted every which way as the ground slumps from the melting permafrost. The thawed remains of melting permafrost also produce many topographic changes. The increased presence of thermokarst lakes and caved-in shorelines provides visible clues of thawing permafrost. These changes in ground-surface stability also pose significant problems for development in Arctic regions. In some areas, the upper layers of permafrost are melting at an alarming rate of 20cm per year (Strom 2004). Many builders were not aware of global warming 30 years ago and therefore did not anticipate the accelerated rates of melting permafrost and its destabilizing effects. Consequently, homes and commercial buildings residing in regions with melting permafrost are now collapsing. Moreover, major roads and airport runways are experiencing dangerous failures as road surfaces are buckling and fracturing. Likewise, the environmental impacts of failing oil or natural gas pipelines in permafrost areas can be high. Sustainable development in periglacial environments will therefore require ways to adapt to thawing permafrost as well as techniques to help retrofit existing structures.

## Solutions: Applying the Precautionary Principle

The Earth is warming. These changes are already manifesting themselves in a series of cascading events that are starting to have profound consequences for humans. Events such as increased flooding, drought, sea-level rise and frequency of strong hurricanes could potentially have devastating effects on human health, food production and the economy. The way in which these consequences play out is up in the air. The future of global warming is in large part dependent upon how humans respond to it.

Ultimately there are two basic responses to global warming in the face of scientific uncertainty. One response would be to do nothing to combat global warming. Proponents of this idea argue that there is not sufficient evidence that human activity is causing global warming and it is unlikely that global warming will result in major environmental damage. Instead of taking action, proponents of this stance insist that it would be more cost effective if the world population live and adapt to future climate change. The limitations to this entirely reactive approach are that surprises and unexpected problems will emerge. For example, while many system changes are slow and predictable, some changes are not quite as linear. As a system approaches a tipping point, changes have a tendency to become magnified, uncontrolled and chaotic. The collapse and melting of the western Antarctica's ice sheet, for instance, would have a devastating impact on millions of coastal inhabitants that would be nearly impossible to respond and adapt to.

The melting of glacial ice should therefore be recognized not just as a sign of global warming, but also as a call to action. This call to action summarizes the second basic response to global warming. According to the **Precautionary Principle**, when faced with scientific uncertainty it is "better to be safe than sorry." This proactive approach to global warming works to mitigate the problem through cost-effective ways of reducing emissions of greenhouse gases and enhancing biospheric sinks for carbon dioxide.

# Cleanup and Prevention Strategies for Reducing Greenhouse Gas Emissions

## Cleanup strategies

Cleanup strategies focus on ways of dealing with greenhouse gas emission after they have been produced. Geologic sequestration of carbon for example, is a cleanup method designed to reduce the amount of carbon dioxide in the atmosphere. **Carbon sequestration** involves capturing carbon dioxide from power plants and industrial smokestacks and injecting it into deep subsurface geologic reservoirs composed of sedimentary rocks. Such carbon capture and storage projects are in progress beneath the North Sea. Natural-gas production facilities in Norway have injected approximately one million tons of carbon dioxide 1,000m deep, below a natural gas field every year since 1996 (Botkin and Keller 2005). While this technological fix appears promising, carbon capture and storage has its pitfalls. The stored carbon dioxide needs to remain sealed from the atmosphere forever. Leaks both small and large caused by earthquake activity could potentially accelerate atmospheric warming in a short period of time. Moreover, carbon sequestration is extremely energy intensive. Many analysts fear that this cleanup strategy may actually increase the use of fossil fuels such as coal and argue that prevention approaches may prove more effective over the long run.

Other sequestration methods include planting large areas of degraded land with fast growing trees and perennial plants. As noted earlier, plants are important carbon sinks that remove carbon dioxide from the air and store it in the soil and their biomass.

## Prevention strategies

Prevention is almost always more effective than cleanup. Tyler Miller and Scott Spoolman point to a study conducted by the US National Academy of Sciences that identified four prevention strategies that should be employed in order to slow atmospheric warming. These strategies include improving energy efficiency, switching to renewable energy resources, preventing deforestation and shifting to a more sustainable form of agriculture. According to this study, these prevention strategies could potentially reduce greenhouse gas emissions by 57–83 percent by the year 2050 (Miller and Spoolman 2012, 513).

It is paramount that not one but all prevention strategies are implemented so that major emission reductions take place in the near future. While a few skeptics insist that these prevention strategies will stall economic growth, numerous scientists and economists argue that these actions will lead to many other environmental, health, political and economic benefits. Improving energy efficiency, for example, reduces oil imports and improves energy security. Likewise, switching from non-renewable energy resources to renewable energy resources creates jobs and decreases air and water pollution by minimizing the combustion of fossil fuels. Moreover, curbing the destruction of the world's tropical forests helps to ensure the preservation of the world's ecosystems and threatened biodiversity. Each of the aforementioned strategies, therefore, are not only effective approaches to stalling global warming, but are simply good policy.

## International Climate Negotiations

Solving the problems of global warming and reducing the emissions of greenhouse gases will of course require a global response. This global response began in 1992 at the Earth Summit in Rio de Janeiro, Brazil. Over 160 countries signed onto a strategy for action to slow and stabilize climate change. While the agreements were not legally binding, it did provide a blueprint for the reduction of global emissions. Symbolically, the Earth Summit was in part a success because it provided an international platform on which the realities of global warming were first recognized.

Five years later, 2,200 delegates from 161 countries convened in Kyoto, Japan in an effort to implement legally binding emission limits. After much deliberation, 187 of the world's 194 countries ratified an agreement that became known as the **Kyoto Protocol**. The first phase of the resulting Kyoto Protocol went into effect in 2005 and required 36 of the more developed countries to cut their greenhouse gas emissions to an average of 5.2 percent below 1990 levels by the year 2012. To assist in emissions reduction, the Kyoto Protocol included a variety of mechanisms. A central component of the Kyoto Protocol is an innovative market-based solution known as **carbon trading** or a cap and trade policy. Carbon trading works by assigning participating countries limits on total allowable emissions. This allows industrialized countries to purchase carbon permits in the marketplace from countries that find it easier to meet their emission targets. In effect, Houghton contends, this trading scheme provides a powerful incentive to cut emissions and utilize energy-efficient technologies and renewable energy resources (2009, 248).

Under the Kyoto Protocol, developing countries are not required to restrict emissions in order to avoid hindering their economic growth. The Kyoto Protocol therefore assigned the greatest responsibility to the industrialized countries of the world, as they are the largest emitters. Case in point, the United States accounts for 5 percent of the world's population yet it emits close to 20 percent of the atmospheric carbon dioxide. Nevertheless, in 2001, President George W. Bush rejected the Kyoto Protocol because he argued that it would harm the United State's economy by limiting the use of fossil fuels. This rejection was a major disappointment to the United State's European allies and was seen by many environmentalists as a significant setback in the international efforts to curb greenhouse gas emissions.

Although the Kyoto Protocol provided an important first step in the mitigation of climate change through the reductions in greenhouse gases, it did have its limitations. The enforcement mechanisms, for example, were unclear and not strict enough. Likewise, "even if the Kyoto Protocol goals were met," according to Robert Strom, "it would still be woefully inadequate to do anything but barely slow down global warming; it would not come close to reducing emissions sufficiently to stabilize the atmosphere" (2007, 234) Despite the limited success of international climate negotiations, many countries, local governments, companies and individuals are confronting the wide array of challenges of global warming.

## Individuals Matter

Each of us contributes to the acceleration of atmospheric warming. Each of us therefore contributes to the breakup and disintegration of the Earth's cryosphere. For better or for worse, individual choices while seemingly insignificant, accumulate over time. Just as the cumulative effect of individual actions has raised atmospheric concentrations of carbon dioxide, it can just as easily reduce it. There are many positive steps that the public can take to slow global warming and reduce the rate of melting of Earth's glaciers. Individual choices such as driving a fuel-efficient car, using energy-efficient appliances and planting trees are of course active ways to reduce carbon dioxide emissions. For more ideas and inspiration go to **carbonfootprint.com**. This website allows you to calculate the amount of carbon dioxide generated by your lifestyle and identifies ways in which you can reduce your carbon dioxide emissions.

If these strategies do not inspire you, perhaps the most important thing you can do for the welfare of our planet is to visit a glacier. Witnessing a glacier's majesty first hand is a transformative experience. Although extremely informative, vivid pictures and dramatic satellite imagery fail to capture the essence of what it is like to stand in front of or on top of a glacier. The sights and sounds of a calving glacier in particular can quickly translate intellectual interest into emotional concern. Whether you think of Gaia theory as sound science or an attractive metaphor, after visiting a glacier it becomes clear that we are intimately connected to these extraordinary elements of nature. The experience will not only change you to the core, but it might just spur you into action.

# Chapter 12 Review

## Key terms

Atmosphere

Troposphere

Greenhouse effect

Global warming

Precautionary principle

Carbon sequestration

Kyoto protocol

Carbon trading

## Concept review and critical thinking

1. Describe the atmosphere's structure and composition.

2. What is the greenhouse effect and what is its importance to global climate change? Identify the major greenhouse gases and describe the human activities that produce them.

3. What are the major negative and positive feedback cycles that might affect the rate of atmospheric warming?

4. What are the consequences of global warming? Include in your discussion an analysis of the impacts on alpine glaciers, continental glaciers, global sea levels and periglacial environments.

5. Should the Precautionary Principle be applied to global warming? Explain.

6. What was the purpose of the Kyoto Protocol? Explain carbon trading.

## Google Earth analysis

The Sexton Glacier is located in Montana's Glacier National Park. The alpine glacier is situated in a cirque on the southeast slope of Matahpi Peak at an elevation between 2,100m and 2,400m. Like many of the glaciers in the park, the Sexton Glacier is receding. According to the United States Geological Survey, small alpine glaciers like the Sexton Glacier are good indicators of climate change. Studies by the survey reveal that the Sexton Glacier has lost over 30 percent of its surface area between 1966 and 2005. Aside from the aesthetic loss of this majestic glacier, the ecological consequences may prove severe. Scientists contend that the accelerated glacial recession may affect the park's ecosystem. For example, a reduction in glacial meltwater could potentially raise stream temperatures and cause the local extinction of temperature-sensitive aquatic species.

## Instructions

Take a trip to Glacier National Park's Sexton Glacier using Google Earth. Type in the latitude/longitude coordinates found below into the "Fly To" box.

*Sexton Glacier: 48.701107, −113.635365*

Unfortunately, the free version of Google Earth does not allow the user to calculate surface area. Instead we will determine the perimeter of the Sexton Glacier using the "Ruler Tool."

1. Determine in meters the perimeter of the Sexton Glacier. Click on the "Ruler Tool" and select the "Path" tab.

2. To determine the perimeter of the glacier, simply left click along the outer edge of the glacier and outline its shape by creating a path that circumnavigates the ice.

3. Using the current glacial-perimeter value, calculate the 1966 perimeter assuming that the current extent is 70 percent of what it was.

# FINAL PROJECT

With Google Earth you can find any location on the Earth and label it with a placemark. By compiling a series of custom placemarks in a folder, you can create, save and play an animated tour of various places on Earth. For the culminating project of this book you will be asked to synthesize your learning by creating and presenting a tour of various glacial landforms that best exemplify the vast array of glacial processes you have learned about.

## Assignment

1. Create a Google Earth tour using the detailed, illustrated instructions provided on the next page.

2. Select your tour theme from the following topics:

   a. *Classifying glaciers.* Create a tour that shows how glaciers are classified. Be sure to include both topographic classifications and temperature classifications. Consider stopping at a variety of glaciers to compare and contrast alpine glaciers with continental glaciers. Your tour should also highlight the differences between highland ice fields, piedmont glaciers and tidewater glaciers.

   b. *Erosional landforms.* Create a tour that shows how glacial erosion creates large-scale features in the landscape. Research and locate erosional landforms such as cirques, arêtes, horns, troughs and paternoster lakes. Be sure to explain how each of these features formed.

   c. *Depositional landforms.* Create a tour that shows how glacial deposition creates large-scale features in the landscape. Research and locate depositional landforms such as terminal moraines, lateral moraines, medial moraines, eskers, kames, kettles and drumlins. Be sure to explain how each of these features formed.

   d. *Global warming and glaciers.* Create a tour that shows how global warming is affecting both continental and alpine glaciers. Research and locate specific glaciers that are currently receding and are particularly susceptible to atmospheric warming. Explain the consequences of such warming and highlight locations on Earth that are especially vulnerable to outcomes such as rising sea level.

3. Provide multiple views of your landforms – change the altitude, angle and direction of view etc. (minimum of four perspectives per landform).

4. The timing of the tour should last for a minimum of 6 minutes and a maximum of 10 minutes.

## Learn to Create a Tour in Google Earth

1. Create a tour folder. Right click on "My Places" located in the "Places" column and select "Add" and then select "Folder." Label the new folder with an appropriate title.

**Figure 37.** Creating a new folder in Google Earth

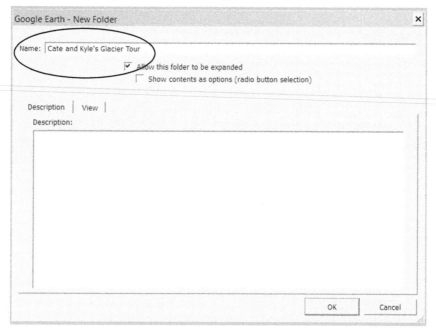

Source: Google Earth is a trademark of Google, Inc.

2. Add placemarks to your new tour folder. Click to highlight your tour folder and navigate to your desired location on the globe. Add a new placemark by clicking on the "Add Placemark" icon located in the tool bar above the Google Earth window.

**Figure 38.** Adding a placemark in Google Earth

Source: Google Earth is a trademark of Google, Inc.

3. Label the new placemark with an appropriate title. It is helpful to give each placemark a number so that you can easily sequence the placemarks in your tour folder as you see fit.

**Figure 39.** Adding a title to a new placemark in Google Earth

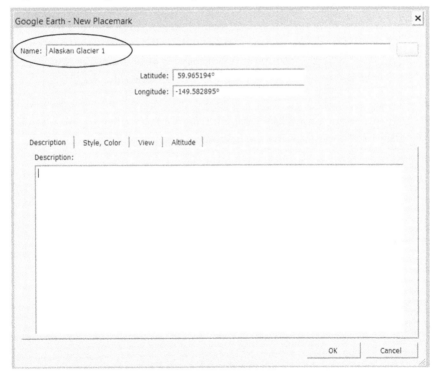

Source: Google Earth is a trademark of Google, Inc.

4. Check to make sure the new placemark is in your tour folder. Notice how the new placemark is slightly indented within the tour folder as shown by the screen shot below. If the new placemark is not located in the tour folder simply click and drag it into your tour folder.

**Figure 40.** Arranging a tour folder in Google Earth

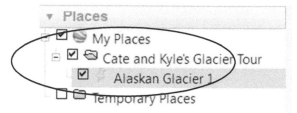

Source: Google Earth is a trademark of Google, Inc.

5. Repeat steps 3 and 4 to add additional placemarks.

Hint! Effective tours often have multiple placemarks with different angles and views at each location. This provides the audience with a variety of perspectives of a particular landform. It is generally more informative to start your placemarks from far away so that the audience has a geographic context for where the location is. Subsequent placemarks can then bring the viewer closer and closer revealing more detail along the way.

6. Playing the tour. Once you are satisfied with your placemarks and their sequence in the tour folder select the tour folder by clicking on it. Select the "Play Tour" icon found at the bottom of the "Places" column.

**Figure 41.** Playing a tour in Google Earth

Source: Google Earth is a trademark of Google, Inc.

7. Saving the tour. Right click on your tour folder and select "Save Place As." Select an appropriate title and a location to save it to.

# BIBLIOGRAPHY

Benn, Douglas I. and David J. A. Evans. 2010. *Glaciers & Glaciation*. 2nd ed. London: Hodder Education.

Bennett, Matthew R. and Neil F. Glasser. 2009. *Glacial Geology: Ice Sheets and Landforms*. 2nd ed. West Sussex: John Wiley & Sons.

Botkin, Daniel B. and Edward A. Keller. 2005 *Environmental Science: Earth as a Living Planet*. 5th ed. Hoboken, NJ: John Wiley & Sons.

Brown, Paul. 1996. *Global Warming: Can Civilization Survive?* London: Blandford.

Colbeck, Samuel C., ed. 1980. *Dynamics of Snow and Ice Masses*. New York: Academic Press.

Davis, Neil. 2001 *Permafrost: A Guide to Frozen Ground in Transition*. Fairbanks: University of Alaska Press.

Eisma, Doeke, ed. 1995. *Climate Change: Impact on Coastal Habitation*. Texel, Netherlands: Lewis Publishers.

Erickson, Jon. 1996. *Glacial Geology: How Ice Shapes the Land*. New York: Facts on File.

French, Hugh M. 1996. *The Periglacial Environment*. 2nd ed. Essex: Addison Wesley Longman.

Hambrey, Michael and Jürg Alean. 2004. *Glaciers*. 2nd ed. Cambridge: Cambridge University Press.

Houghton, John. 2009. *Global Warming: The Complete Briefing*. 4th ed. Cambridge: Cambridge University Press.

IPCC. 2001. *The Intergovernmental Panel on Climate Change Scientific Assessment*. New York: Oxford University Press.

_____. 2007. *4th Assessment Report*. United Nations.

Johannessen, O. M., K. Khvorostovsky, M. W. Miles and L. P. Bobylev. 2005. "Recent ice-sheet growth in the interior of Greenland." *Science* 310 (November): 1013–16.

Joughin, I., W. Abdalati and M. Fahnestock. 2004. "Large fluctuations in speed on Greenland's Jakobshavn glacier." *Nature* 432 (December): 608–10.

Krabill, W., E. Hanna, P. Huybrechts, W. Abdalati, J. Cappelen, B. Csatho, E. Frederick, S. Manizade, C. Martin, J. Sonntag, R. Swift, R. Thomas and Yungel. 2004. "Greenland ice sheet: Increased coastal thinning." *Geophysical Research Letters* 31: L24402.

Lovelock, James. 1979. *Gaia: A New Look at Life on Earth.* Oxford: Oxford University Press.

Martens, Pim and Jan Rotmans, eds. 1999. *Climate Change: An Integrated Perspective.* Dordrecht: Kluwer Academic Publishers.

Menzies, John, ed. 1995. *Modern Glacial Environments: Processes, Dynamics and Sediments,* vol. 1. Oxford: Butterworth-Heinemann.

_____, ed. 2002. *Modern and Past Glacial Environments.* Oxford: Butterworth-Heinemann.

Miller, Tyler and Scott Spoolman. 2012. *Living in the Environment.* 17th ed. Belmont, CA: Brooks/Cole.

Muir, John. 1989 [1915]. *Travels in Alaska.* San Francisco: Sierra Club.

Sharp, Robert P. 1988. *Living Ice: Understanding Glaciers and Glaciation.* Cambridge: Cambridge University Press.

Shepherd, A., D. Wingham and E. Rignot. 2004. "Warm ocean is eroding west Antarctic ice sheet." *Geophysical Research Letters* 31: L23402.

Strom, Robert. 2007. *Hot House: Global Climate Change and the Human Condition.* New York: Copernicus Books.

Sugden, David E. and Brian S. John. 1976. *Glaciers and Landscape: A Geomorphological Approach.* London: Edward Arnold.

Street, R. B. and P. I. Melnikov. 1990. "Seasonal snow cover, ice and permafrost." In *Climate Change: The IPCC Impacts Assessment,* ch. 7, sections 7-1 to 7-33. Report prepared for IPCC by Working Group II. Canberra: WMO-UNEP, Australian Government Publishing Service.

Washburn, A. L. 1973. *Periglacial Processes and Environments.* New York: St Martin's Press.

Williams, Peter J. and Michael W. Smith. 1989 *The Frozen Earth: Fundamentals of Geocryology.* Cambridge: Cambridge University Press.

# INDEX

## A

ablation 9, 11–13, 25, 29, 33
abrasion 38–41, 43–4, 49
accordant junctions 46
accumulation 9, 13, 25, 29
active layer 77, 79, 83
Alaska 5, 13, 15–16, 35–6, 46, 68, 98
Alaska's North Slope 80–81
albedo 11–12, 91
Alberta (Canada) 58
Alean, Jürg 31
Aletsch Glacier 27
alpine glaciers 2, 15–17, 20–21, 25–7, 34, 38, 40, 92–3, 98, 104
Amazon River 88
Andes 4, 93
Antarctic 19, 46, 92–3
Antarctica 3, 18, 66, 68, 97–8
Antarctic ice sheet 12, 95, 97–9
Antarctic Peninsula xv–xvi, 3, 16, 85, 95
aquifers 4, 96
Arctic 19, 46, 68, 89, 92, 94, 99
Arctic Circle 18, 93
arêtes 47–8, 50
Atlantic Ocean 71–3
atmosphere 1, 6, 65–8, 85–7, 89–91, 97–8, 100, 102, 103
Austria 4
avalanches 37–8
axial tilt 64–5

## B

Baffin Island (Canada) 19
Bangladesh 96
basal debris 39, 41
basal melting 19, 54, 94
basal meltwater 38, 40–41
basal sliding (slipping) 23–6, 34
basalt 71
Benn, Douglas 15, 23, 29, 44, 64, 66, 70
Bennett, Mathew 23–4, 37, 46, 55, 57
Bernese Alps 27
biodiversity 4
biosphere 1, 86
block weakening 39, 40
blue ice 10
Bolivia 4, 93
boreal forests: see taiga forest
Boston (Massachusetts) 96
braided streams 55–6
British Isles 69
Bush, George W. 102

## C

Cadair Idris Valley
calcium carbonate 72
Calgary (Canada) 55
calving 11–12, 18
Canada 68, 98
Canadian Arctic 16, 68, 77
Cape Cod 58

carbonate rocks 87
carbon cycle 86–7
carbon dioxide 2, 66, 86–7, 89–90, 97–8, 100
carbon sequestration 100, 103
carbon sinks 87, 100
carbon trading 101, 103
chattermarks 39
cirques 16, 47, 50
Clew Bay (Ireland) 60
climate change 5, 63–74, 85–104
cloud-radiation feedback 89–90
cold glaciers 19–20, 26
compressing flow 25, 27
Concordia 27
continental glacier 15, 17–21, 38, 68, 73
Cordilleran ice sheet 68
creep 23, 26, 31, 37
crescentic gouges 39
crevasses 23, 30–33, 35, 38
cryogenic weathering 76, 81, 83
cryosphere 2–3, 6, 75 , 85, 91, 102
cryosuction 76, 83
cryoturbation 83

**D**

Davis, Neil 77, 83
deep-sea cores 67
deforestation 86–8, 101
deposition
    direct deposition 53–6
    indirect deposition 53, 55–6
dilatation joints 39–40
discordant junctions 46
dolostone 72
drumlins 59–62, 72

**E**

Earth Summit in Rio de Janeiro, Brazil 101
Easton Glacier 30
Ecuador 93
Ellesmere Baffin glacier complex 68
Ellesmere Island 19
El Niño 91
englacial debris 38, 41

equilibrium line 9, 11–13
EPICA: see European Project for Ice Coring in Antarctica (EPICA)
Erickson, Jon 82
erosion 37–51, 72, 76
eskers 60–62, 72
Eurasia 68
Europe 4, 65, 69, 97
European Alps 38, 69
European Project for Ice Coring in Antarctica (EPICA) 66
Evans, David 15, 23, 29, 44, 64, 66, 70
extending ice flow 25, 27

**F**

faulting 33–5
feedback loops
    negative feedback loops 2, 6, 91
    positive feedback loop 2, 6, 89–90, 98
Finger Lakes 4–5
fjords 5, 15, 44, 46–8, 50
folding 24, 33–5
foliation 33–5
fossil fuels 5, 86–7, 89, 102
fracture 23–4, 26, 31
French, Hugh M. 75, 78
frost action 38–9, 75–7, 82–3
frost heaving 75–6, 83–4

**G**

Gaia theory 1, 6, 103
Gangotri Glacier 91
Garden Wall 48
Geiranger Fjord 47
Gilkey Glacier 35–6
glacial
    deposition (see deposition)
    erosion (see erosion)
    erratics 54–6, 72
    firn 9, 12
    flour 39, 41
    meltwater 4, 20, 24, 39, 55, 61, 93–4
    periods 4, 63
    striations 39, 41, 43
    tarn 49, 50

till 53–6, 72
troughs 44–6, 50
Glacier Bay 5–7
glacierets 17, 20
Glacier National Park 3, 104
glacifluvial material 55–6, 60–61
glacio-eustasy 70, 73–4
glacio-isostasy 70, 73–4
Glasser, Neil 23–4, 37, 46, 55, 57
global warming xv, 3, 77, 85–104, 106
Google Earth xvi–xx, 7, 13, 21, 27, 36, 51, 62, 104–8
gravity 37, 57
Great Lakes 4, 70–74
  Lake Erie 71–2
  Lake Huron 71–2
  Lake Iroquois 72–3
  Lake Michigan 71–2
  Lake Ontario 71–2
  Lake Superior 71–2
greenhouse effect 65–6, 73, 86, 103
greenhouse gasses 2, 66, 86–7, 89, 98, 100–103
Greenland 3, 15–16, 18, 66, 91, 93–5, 97–8
groundwater 4

**H**

Hambrey, Michael 31
hanging valley 5–6, 46, 50
High Arctic archipelago (Svalbard) 19
highland icefield 16, 20
Himalayans 32, 38
horizontal sorting 55–6
horns 47, 49–50
Houghton, Sir John 89, 102
hydroelectric power 4, 93
hydrosphere 1, 6, 86

**I**

ice-albedo feedback 89, 91, 94
ice ages 3, 63–73
ice aprons 17, 20
icecaps 2, 16, 19
ice core analysis 66–7, 96
ice dams 73
icefalls 29–32, 36

ice fringes 17, 20
Iceland 19
ice sheets 2–3, 16–18, 58, 63, 68–70, 72–3, 94–8
ice shelves 18, 94–6
ice wedges 76, 83
Illinois 69
Indonesia 65
infrared energy 11
interglacial periods 3, 63
Intergovernmental Panel on Climate Change (IPCC) 89, 94
internal deformation 23, 25–6, 31
IPCC: see Intergovernmental Panel on Climate Change (IPCC)
irrigation 4, 93
isostatic rebound 73

**J**

John, Brian S. 24, 40
Juneau Icefield 13, 35

**K**

kames 61–2
kettles 61–2
Khumbu Icefall 31–2
Kyoto Protocol 101–3

**L**

Lake Erie 71–2
Lake Huron 71–2
Lake Iroquois 72–3
Lake Michigan 71–2
Lake Ontario 71–2
Lake Superior 71–2
landforms of glacial deposition
  ice marginal features 57–9, 61
  ice contact features 60–62
  subglacial features 57, 59–61
landforms of glacial erosion
  intermediate scale features of glacial erosion 43–4, 50
  large-scale landforms of glacial erosion 43–50
landslides 37

Laurentide ice sheet 68–9, 73
law of superposition 33, 35
Larsen-B ice shelf xv–xvi, 85, 95–6
lateral moraines 58–9, 61
Lembert Dome 54
limestone 72
lithosphere 1, 6, 65, 86
"Little Ice Age" 64
lodgement till 54, 56
Long Island 58, 69
longitudinal bars 56
longitudinal movement 25–6
Lovelock, James 1

**M**

mass balance 9, 11–13, 95
mass movements 37, 58
Matahpi Peak 104
Matterhorn 49–50
medial moraines 58–9, 61
melt-out till 54, 56
Mendenhall Glacier 13
Menzies, John 5
mesosphere 85
metamorphic rocks 33–4
methane 66, 86–7, 89, 98
Milankovitch cycles 64, 73
Mississippi River 72
moraines 57–9, 61, 72
moulins 94
mountain glaciers: see alpine glaciers
Mount Baker 45
Mount Everest 16
Mount Kilimanjaro 3
Mount Pinatubu 74
mudflows 37
Muir, John 5–7

**N**

New England 60
New York 60
New York City 96
New Zealand 46
Niagara Falls 72–3
nitrous oxide 66, 86, 89
North America 4, 58, 60, 65, 68–72, 97

North Slope: see Alaska's North Slope
Norway 4, 46, 92
nunataks 16, 20

**O**

ocean-circulation feedback 89–90
ocean currents 4
Ogallala Aquifer 4
ogives 29–30, 35–6
    band ogives 29–30, 35–6
    wave ogives 29–30, 35–6
orbital shape 64
outwash plains 55–6
oxygen-isotope analysis 67

**P**

Patagonia 3, 16, 48, 59
Pangaea 66
patterned ground 81–4
periglacial 75, 79–84, 98–9
permafrost 2, 75, 77–81, 83–4, 89, 98–9
Peru 4, 93
Petermann Glacier 93
Philippines 74
photosynthesis 86
piedmont glaciers 16, 20
pingos 79–80, 83
plate tectonics 64–5
Pleistocene Epoch 66, 68–73, 75, 77
plucking 38–41, 43–4, 49
pollen evidence 67
polythermal glaciers 20, 34, 39
Precautionary Principle 99, 103
pressure unloading 39

**Q**

quarrying: see plucking

**R**

recessional moraines 57–8, 61
renewable energy 4
respiration 86
Rhone Valley 27
rift valley 71–2

roches moutonnées 43–4, 50
Rochester (New York) 73
rock cycle 33–4
Rodina 66

## S

Scandinavia 98
sea-level change 4, 70, 96–8
sedimentary rocks 33–4, 72
seracs 32, 35
Sexton Glacier 104
Sharp, Robert 54, 61
Siberia 78, 98
solar irradiance 63–4
solar radiation 11, 15
South America 46
South Georgia Island 17
South Pole 19
snowline 15, 20–21
St Lawrence River 71
St Lawrence Seaway 72–3
stratosphere 85
Strom, Robert 64, 102
subglacial bed deformation 23, 26
subglacial debris 37–8, 41
Sugden, David E 24, 40
subsurface ice structures 29, 32–5
sunspot cycles 64
supraglacial debris 37–8, 41
surface ice structures 29–32, 35
Svalbard Archipelago: see High Arctic
    archipelago (Svalbard)
Swiss Alps 4–5, 26, 27, 46, 50
Switzerland 4, 92
system 1–2, 6

## T

taiga forest 78
taliks 79, 83
Tambora volcano 65
terminal moraines 57–8, 61

terminus 7, 13, 29, 35
thaw weakening 75–6, 83–4
thermal expansion 97
thermokarst 80–81, 83–4, 99
thermosphere 85
thrust faults 34–5
tidewater glacier 15, 17, 20
tipping point 91, 99
transverse movement 25–7, 36
troposphere 85–6, 103
Tuktoyaktuk (Canada) 80

## U

unconformities 33, 35
United States 54, 58, 68, 96, 102
Upsala Glacier 59
U-shaped valley 6, 26, 44–6

## V

valley glaciers: see alpine glaciers
viscoplastic material 23
volcanoes 65–6, 71, 74, 91

## W

warm glaciers 19–20, 26
Washburn, A. L. 83
wastage 11–13, 33
water cycle 4
water vapor 11, 85–6, 89–90, 94
water-vapor feedback 89
weathering 39, 66
whalebacks 44, 50
Wisconsinan Stage 69–70, 77
World Glacier Monitoring Service 93

## Y

Yosemite Valley 6–7, 54
Yukon 78

Milton Keynes UK
Ingram Content Group UK Ltd.
UKHW040617240924
448754UK00001B/20

9 780857 280619